This series aims to report new developments in physical research and teaching – quickly, informally, and at a high level. The type of material considered for publication includes:

1. Preliminary drafts of original papers and monographs

2. Lectures on a new field, or presenting a new angle on a classical field

3. collections of seminar papers

4. Reports of meetings

Texts which are out of print but still in demand may also be considered if they fall within these categories.

The timeliness of a manuscript is more important than its form, which may be unfinished or tentative. Thus, in some instances, proofs may be merely outlined and results presented which have been or will later be published elsewhere.

Publication of *Lecture Notes* is intended as a service to the international physical community, in that a commercial publisher, Springer-Verlag, can offer a wider distribution to documents which would otherwise have a restricted readership. Once published and copyrighted, they can be documented in the scientific libraries.

Manuscripts

Manuscripts are reproduced by a photographic process; they must therefore be typed with extreme care. Symbols not on the typewriter should be inserted by hand in indelible black ink. Corrections to the typescript should be made by sticking the amended text over the old one, or by obliterating errors with white correcting fluid. The figures (in the original size) ready for reproduction should be inserted into the text. Should the text, or any part of it, have to be retyped, the author will be reimbursed upon publication of the volume. Authors receive 50 free copies.

The typescript is reduced slightly in size during reproduction, therefore a large size of type should be used; best results will not be obtained unless the text on any one page is kept within the overall limit of 18 x 26.5 cm (7 x 10½ inches). **The publishers will be pleased to supply on request special stationery with the typing area outlined.**

Manuscripts in English, German or French should be sent to Springer-Verlag, 6900 Heidelberg, Postfach 1780.

Die „*Lecture Notes*" sollen rasch und informell, aber auf hohem Niveau, über neue Entwicklungen in der Physik berichten. Zur Veröffentlichung kommen:

1. Vorläufige Fassungen von Originalarbeiten und Monographien.

2. Spezielle Vorlesungen über ein neues Gebiet oder ein klassisches Gebiet in neuer Betrachtungsweise.

3. Seminarausarbeitungen.

4. Vorträge von Tagungen.

Ferner kommen auch ältere vergriffene spezielle Vorlesungen, Seminare und Berichte in Frage, wenn nach ihnen eine anhaltende Nachfrage besteht.

Die Beiträge dürfen im Interesse einer größeren Aktualität durchaus den Charakter des Unfertigen und Vorläufigen haben. Sie brauchen Beweise unter Umständen nur zu skizzieren und dürfen auch Ergebnisse enthalten, die in ähnlicher Form schon erschienen sind oder später erscheinen sollen.

Die Herausgabe der „*Lecture Notes*" Serie durch den Springer-Verlag stellt eine Dienstleistung an die physikalischen Institute dar, indem der Springer-Verlag für ausreichende Lagerhaltung sorgt und einen großen internationalen Kreis von Interessenten erfassen kann. Durch Anzeigen in Fachzeitschriften, Aufnahme in Kataloge und durch Anmeldung zum Copyright sowie durch die Versendung von Besprechungsexemplaren wird eine lückenlose Dokumentation in den wissenschaftlichen Bibliotheken ermöglicht.

Lecture Notes in Physics

9

Derek W. Robinson
Université d'Aix-Marseille II
Centre de Luminy

The Thermodynamic Pressure in Quantum Statistical Mechanics

Springer-Verlag Berlin Heidelberg GmbH 1971

ISBN 978-3-540-05640-9 ISBN 978-3-540-36914-1 (eBook)
DOI 10.1007/978-3-540-36914-1

Originally published by Springer-Verlag Berlin Heidelberg New York in 1971.

Library of Congress Catalog Card Number 72-175913.

Offsetdruck: Julius Beltz, Hemsbach/Bergstr.

PREFACE

The following notes represent a course of lectures delivered at the University of Lausanne in October and November 1970 as part of the programme of the 'troisième cycle en Suisse Romande'. My participation in this programme was instigated and arranged by Prof. M. Guenin.

The first half of this course is an expanded and revised version of lectures held at the University of Marseilles in Spring 1970 and is partially based upon the 'troisième cycle' theses of J.L. Maltret and P. Rolot.

D.W. ROBINSON

Bandol, November 1970

C O N T E N T S

C O N T E N T S

CHAPTER 0

GENERAL INTRODUCTION

In these lectures we examine and derive various properties of the thermodynamic pressure of point particles in Quantum Statistical Mechanics. Our main aim is to demonstrate that the pressure can be calculated from a variational principle ; the thermodynamic pressure is given by the supremum over a class of invariant states of the entropy per unit volume at fixed energy per unit volume. This type of variational principle has been discussed previously for lattice and hard core systems, both classically and quantum mechanically. It is of principal interest because it can provide information concerning the equilibrium states of the system in addition to properties of the pressure. For example in the simpler models it has been used to derive conditions under which the system is in a pure phase or, alternatively, in a mixed phase.

The derivation of the variational principle for quantum mechanical point particles presents a number of difficulties which are not present for classical systems or quantum lattice systems. The major problem which occurs is connected with the fact that the pressure can be defined with various different boundary conditions. To establish the desired variational principle it is necessary to derive estimates which are sufficient to show that the pressure is independent of the boundary conditions used in its definition. Thus in the first half of these lectures we will devote our consideration to the dependence of the pressure on various forms of boundary conditions. In the second half we consider the construction of the variational principle.

We have attempted to make these lectures self-contained by summarizing in each chapter the mathematical definitions and results which we use. Due to restrictions of time and space it was not possible to give derivations of these standard results. We give however references at the end of each chapter which are intended for general information relevant to the content of the chapter. We have not attempted to list all references related to the research problems we discuss and we apologise to anyone we might have offended by this negligent practise In the same spirit we have not emphasized in any way the new results which appear in the lectures. We admit at the outset that these results are to a large extent unsatisfactory, prinpally because of the strong restrictions we place upon the inter-particle interactions. We consider the present description only as a zero-order presentation of the properties of the pressure and hope that these notes will serve as an introduction to the many problems which remain open.

CHAPTER I

THE HAMILTONIANS OF FINITE SYSTEMS

INTRODUCTION

The description of quantum mechanical systems is given by the kinematical specification of a Hilbert space $\mathcal{H}$ of vector states of the system and the dynamical specification of the Hamiltonian as a self-adjoint operator acting on $\mathcal{H}$. The mathematical requirement of self-adjointness corresponds physically to the conservation of probability and ensures that the dynamics are completely specified. If one wishes to describe a finite system, e.g. a number of particles contained in a closed vessel, the specification of the Hamiltonian involves both the inter-particle interaction and the interactions of the particles with the walls of the container. These two points are essentially independent and in this chapter we concentrate mainly on the difficulties and details of the latter. Thus to a large extent we will consider a finite system of mutually non-interacting particles and study the relation between the specification of boundary conditions, i.e. the interactions of the particles with the walls of the container, and the self-adjoint differential operators suited to the description of "free" particles.

In section §1 we recall a number of standard definitions and results concerning unbounded operators in Hilbert space and illustrate some of the material by the discussion of simple differential operators. In section §2 we review the connection between positive sesquilinear forms over a Hilbert space $\mathcal{H}$ and positive self-adjoint operators acting on $\mathcal{H}$; this material is again illustrated by the discussion of differential operators. Finally in section §3 we apply this preparatory material to the description of the various self-adjoint Hamiltonians of finite systems which will interest us in the sequel ; this discussion includes the operators corresponding to elastic boundary conditions and periodic boundary conditions.

The discussion of sections §1 and §2 is standard mathematical material and is presented practically without proofs with the exception of the illustrative examples of differential operators. Our choice of material is rather selective and is not meant as a complete survey.

§1 - UNBOUNDED OPERATORS ON HILBERT SPACE

1.1.1. A complex pre-Hilbert space is a vector space over the complex numbers $\mathbb{C}$ equipped with a scalar product. Such a space has a unique minimum completion and the completed space is a complex Hilbert space. We use throughout the symbols $\mathcal{H}$, $\mathcal{H}_1$, $\mathcal{H}_2$, ... to denote complex Hilbert spaces and $(\cdot\,,\,\cdot)$, $(\cdot\,,\,\cdot)_1$, ... for the associated inner products. We use the physicist's convention that (ψ, φ) is linear in φ and anti-linear in ψ. We denote the norms of vectors by $\|\psi\|$, $\|\psi\|_1$ etc., i.e.

$$\|\psi\| = \sqrt{(\psi, \psi)} \qquad \psi \in \mathcal{H}$$

Note that as $\mathcal{H}$ is complex the scalar product is obtainable from the norm by the polarization formula

$$(\psi, \varphi) = \frac{1}{4}\left\{\|\varphi+\psi\|^2 - \|\varphi-\psi\|^2 + i\|\varphi+i\psi\|^2 - i\|\varphi-i\psi\|^2\right\}.$$

1.1.2. Let $\{\mathcal{H}_n\}_{n\geq 1}$ be a family of complex Hilbert spaces. The direct sum $\mathcal{H}$ of the $\mathcal{H}_n$ is designated by

$$\mathcal{H} = \bigoplus_{n\geq 1} \mathcal{H}_n = \mathcal{H}_1 \oplus \mathcal{H}_2 \oplus \cdots$$

and consists of the families $\{\psi_n\}_{n\geq 1}$, with $\psi_n \in \mathcal{H}_n$ such that

$$\sum_{n\geq 1} \|\psi_n\|_n^2 < +\infty$$

The scalar product on $\mathcal{H}$ is defined by

$$(\varphi, \psi) = \sum_{n\geq 1} (\varphi_n, \psi_n)_n$$

where $\varphi = \{\varphi_n\}_{n\geq 1}$ and $\psi = \{\psi_n\}_{n\geq 1}$.

Using the above definitions one readily establishes that the direct sum $\mathcal{H}$ is a Hilbert space, in particular it is complete.

1.1.3. Let $\mathcal{H}_1$ and $\mathcal{H}_2$ be two complex Hilbert spaces. Their incomplete tensor product is a pre-Hilbert space specified by the scalar product

$$(\varphi_1 \otimes \varphi_2, \psi_1 \otimes \psi_2) = (\varphi_1, \psi_1)_1 \ (\varphi_2, \psi_2)_2$$

for $\varphi_1, \psi_1 \in \mathcal{H}_1$ and $\varphi_2, \psi_2 \in \mathcal{H}_2$. The unique minimum completion of this pre-Hilbert space is called the tensor product of $\mathcal{H}_1$ and $\mathcal{H}_2$ and is denoted by $\mathcal{H}_1 \otimes \mathcal{H}_2$.

1.1.4. An operator T on the complex Hilbert space $\mathcal{H}$ is a linear subspace $D(T) \subset \mathcal{H}$ and a linear map $T;\ D(T) \to \mathcal{H}$; $D(T)$ is the domain and $TD(T)$ the range of T .

Explicitly the action of T is

$$\psi \in D(T) \longrightarrow T\psi \in TD(T) = R(T)$$

and if $\psi_1, \psi_2 \in D(T)$ and $\alpha_1, \alpha_2 \in \mathbb{C}$ then $\alpha_1\psi_1 + \alpha_2\psi_2 \in D(T)$ and

$$T(\alpha_1\psi_1 + \alpha_2\psi_2) = \alpha_1 T\psi_1 + \alpha_2 T\psi_2 \ .$$

The domain $D(T)$ always contains the zero element 0 and one always has $T0 = 0$

The operator T is said to be densely defined if $D(T)$ is a dense subset of $\mathcal{H}$.

Two operators T_1 and T_2 are regarded as equal if and only if $D(T_1) = D(T_2)$ and $T_1\psi = T_2\psi$ for all $\psi \in D(T_1)$. If on the other hand $D(T_1) \supset D(T_2)$ and $T_1\psi = T_2\psi$ for all $\psi \in D(T_2)$ then T_1 is said to be an extension of T_2 ($T_1 \supset T_2$) and conversely T_2 is a restriction of T_1 ($T_2 \subset T_1$).

The sum $T_1 + T_2$ of two operators T_1 and T_2 is defined by

$D(T_1+T_2) = D(T_1) \cap D(T_2)$ and

$$(T_1 + T_2)\psi = T_1\psi + T_2\psi$$

for $\psi \in D(T_1+T_2)$; the product T_1T_2 is defined by

$$T_1T_2\psi = T_1(T_2\psi)$$

with the domain $D(T_1T_2)$ specified as the $\psi \in D(T_2)$ such that $T_2\psi \in D(T_1)$.

1.1.5. Let T be an operator on $\mathcal{H}$. A sequence $\{\psi_n\}_{n\geq 1}$ of vectors in $D(T)$ is said to be T-convergent (to $\psi \in \mathcal{H}$) if for each $\epsilon > 0$ there is an n_ϵ such that

$$\|\psi_n - \psi_m\| < \epsilon$$

and

$$\|T\psi_n - T\psi_m\| < \epsilon$$

for all $n, m > n_\epsilon$ (and ψ_n converges strongly to ψ).

The operator T is defined to be closed if the T-convergence of $\{\psi_n\}_{n\geq 1}$ to ψ implies both $\psi \in D(T)$ and

$$\lim_{n\to\infty} \| T\psi_n - T\psi \| = 0 .$$

An operator is said to be closable if and only if it has a closed extension or, equivalently, if and only if the conditions

$$\psi_n \in D(T)$$

$$\lim_{n\to\infty} \| \psi_n \| = 0$$

and

$$\lim_{n\to\infty} \| T\psi_n - \psi \| = 0$$

imply that $\psi = 0$. When T is closable there is a closed extension $\tilde{T}$ of T, called the closure of T, which is the smallest closed extension of T in the sense that any closed extension of T is also an extension of $\tilde{T}$. The closure $\tilde{T}$ is defined as follows. The vector $\psi \in D(\tilde{T})$ if and only if there exists a sequence ψ_n which is T-convergent to ψ ; in this case $\tilde{T}\psi$ is defined by

$$\lim_{n\to\infty} \| T\psi_n - \tilde{T}\psi \| = 0 .$$

(The closability condition ensures that $\tilde{T}$ is an operator ; in particular it ensures that $\tilde{T}0 = 0$.)

Let T be a closed operator and S any closable operator such that $\tilde{S} = T$; the domain $D(S)$ of S is called a core of T.

1.1.6. Two operators T and S on $\mathcal{H}$ are said to be adjoint to one another if

$$(\varphi, T\psi) = (S\varphi, \psi)$$

for all $\psi \in D(T)$ and all $\varphi \in D(S)$. Each operator has in general many adjoints but if T is densely defined then there is a unique operator T^* adjoint to T, called the adjoint of T, which is maximal in the sense that T^* is an extension of each operator adjoint to T. T^* is constructed as follows. $D(T^*)$ consists of all $\psi \in \mathcal{H}$ such that there exists a $\varphi \in \mathcal{H}$ with the property that

$$(\psi, T\chi) = (\varphi, \chi)$$

for all $\chi \in D(T)$. (Note that as $D(T)$ is assumed to be dense the φ is uniquely determined by the ψ). The action of T^* on ψ is then defined by

$$T^*\psi = \varphi .$$

<u>The adjoint T^* of a densely defined operator T is always closed.</u> To deduce this consider a sequence $\psi_n \in D(T^*)$ which is T^*-convergent to ψ and denote the limit of $T^*\psi_n$ by φ. One then has for all $\chi \in D(T)$

$$(\psi, T\chi) = \lim_{n\to\infty} (\psi_n, T\chi) = \lim_{n\to\infty} (T^*\psi_n, \chi) = (\varphi, \chi).$$

Thus $\psi \in D(T^*)$ and

$$\lim_{n\to\infty} \| T^*\psi_n - T^*\psi \| = 0$$

i.e. T^* is closed.

More generally one has the following. <u>If the operator T is densely defined and closable, then T^* is densely defined (and closed). Hence $T^{**} = (T^*)^*$ exists. Furthermore T^{**} is equal to the closure $\tilde{T}$ of T.</u>

1.1.7. An operator is said to be <u>symmetric</u> if it is densely defined and

$$(T\varphi, \psi) = (\varphi, T\psi)$$

for all $\varphi, \psi \in D(T)$. An alternative equivalent definition is that T is densely defined

$$D(T^*) \supset D(T)$$

and

$$T^*\psi = T\psi$$

for all $\psi \in D(T)$, i.e. T^* is an extension of T.

<u>Each symmetric operator T is closable (T^* is a closed extension of T) and the operator T^{**} coincides with the closure $\tilde{T}$ of T (cf. 1.1.6.).</u>

As T^* is an extension of T it follows that T^{***} is an extension of T^{**} and since T^{**} is the closure of T it follows that the closure of a symmetric operator is symmetric.

1.1.8. If T is a symmetric operator with the property that

$$(\psi, T\psi) \geq 0$$

for all $\psi \in D(T)$ then T is said to be <u>positive</u> (more precisely non-negative).

1.1.9. An operator T is said to be <u>self-adjoint</u> if it is densely defined and

$T^* = T$. A symmetric operator T is said to be essentially self-adjoint if its closure $T^{**} = \tilde{T}$ is self-adjoint (or, equivalently, if its adjoint T^* is symmetric).

In general a symmetric operator can have many (or no) self-adjoint extensions ; a self-adjoint operator has no proper self-adjoint extensions. (If T is self-adjoint and A is a self-adjoint extension of T then T^* is an extension of A^*. But $T=T^*$ and $A=A^*$, and hence T equals A). An essentially self-adjoint operator has a unique self-adjoint extension namely its closure.

1.1.10. A vector $\psi \in \mathcal{H}$ is defined to be an analytic vector for an operator T if $\psi \in D(T^n)$ for all $n = 1, 2, \cdots$ and

$$f(t) = \sum_{n \geq 1} \frac{t^n}{n!} \|T^n \psi\|$$

is an analytic function of t at $t=0$.

A symmetric operator T is essentially self-adjoint whenever $D(T)$ contains a dense set of analytic vectors for T.

A closed symmetric operator T is self-adjoint if, and only if, $D(T)$ contains a dense set of analytic vectors for T (Nelson, E. ; Ann. Math. 70 572 (1959)).

1.1.11. A second useful criterion for self-adjointness is the following.

Let T be a densely defined, closed operator, then T^*T is self-adjoint and $D(T^*T)$ is a core of T.

Each self-adjoint operator has a unique spectral decomposition which we will not describe. We remark however that this decomposition can be used to uniquely define bounded functions of the operator under consideration.

In the sequel we will principally encounter self-adjoint operators whose spectrum is discrete, i.e. operators with the property that there is a complete set of $\psi \in D(T)$ (eigenfunctions of T) such that

$$T\psi = \lambda\psi$$

for some λ (an eigenvalue). The operators we encounter also have the special property that their eigenvalues have finite multiplicity, i.e. the number of mutually orthogonal eigenfunctions corresponding to any fixed eigenvalue is finite.

1.1.12. Let $\{\mathcal{H}_n\}_{n\geq 1}$ be a family of complex Hilbert spaces and $\mathcal{H}$ their direct sum (defined in 1.1.2.). Further let $\{T_n\}_{n\geq 1}$ be a family of operators on the $\mathcal{H}_n$ with domains $D(T_n)$. The direct sum T,

$$T = \bigoplus_{n\geq 1} T_n = T_1 \oplus T_2 \oplus \cdots$$

is defined on $\mathcal{H}$ by

$$D(T) = \{ \psi = \{\psi_n\}_{n\geq 1} \, ; \, \psi_n \in D(T_n) \, , \, \sum_{n\geq 1} \|T_n \psi_n\|^2 + \|\psi_n\|^2 < +\infty \}$$

and

$$T\psi = T_1\psi_1 \oplus T_2\psi_2 \oplus \cdots$$

It follows that T is densely defined, respectively closable, closed, symmetric, positive, essentially self-adjoint, self-adjoint, if and only if each member of the family $\{T_n\}_{n\geq 1}$ has the corresponding property.

As this simple result does not appear in the standard texts we will indicate the proof of a few of the items. For example let us demonstrate that if each T_n is closable then T is closable. Take $\psi^{(m)} = \{\psi_n^{(m)}\}_{n\geq 1} \in D(T)$ and assume

$$\lim_{m\to\infty} \|\psi^{(m)}\| = 0 \, , \quad \lim_{m\to\infty} \|T\psi^{(m)} - \psi\| = 0 \, .$$

We must show that $\psi = \{\psi_n\}_{n\geq 1} = 0$ i.e. that each $\psi_n = 0$. But

$$\|\psi^{(m)}\| \geq \|\psi_n^{(m)}\| \quad \text{and} \quad \|T\psi^{(m)} - \psi\| \geq \|T_n\psi_n^{(m)} - \psi_n\| \, .$$

But we now have

$$\psi_n^{(m)} \in D(T_n) \, , \quad \lim_{m\to\infty} \|\psi_n^{(m)}\| = 0 \, , \quad \lim_{m\to\infty} \|T_n\psi_n^{(m)} - \psi_n\| = 0$$

and thus $\psi_n = 0$, because T_n is assumed closable. Similarly if each T_n is closed then T is closed.

To prove that the essential self-adjointness of each T_n implies that T is essentially self-adjoint one can use the criterion given in 1.1.10. If $\psi_n \, ; \, n = 1, 2, \ldots N$ are analytic vectors for T_n, $n = 1, 2, \ldots N$ then $\psi = \{\psi_n\}_{n\geq 1}$ with $\psi_n = 0$ for $n > N$ is an analytic vector for T. As each T_n has a family of analytic vectors which is dense in $\mathcal{H}_n$ the set of all analytic vectors of T constructed in the foregoing manner is certainly dense in $\mathcal{H}$ and the desired result follows. Similarly if each T_n is self-adjoint then T is closed symmetric and $D(T)$ contains a dense set of analytic vectors. Thus T is self-adjoint by 1.1.10.

All the reverse implications follow by identifying $\mathcal{H}_m$ $(D(T_m))$ as the subspace of $\mathcal{H}$ $(D(T))$ generated by the vectors $\psi = \{\psi_n\}_{n\geq 1}$ with $\psi_n = 0$ unless $n = m$ ($\psi_n = 0$ unless $n = m$ and $\psi_m \in D(T_m)$).

Let $\mathcal{H}_1$ and $\mathcal{H}_2$ be two complex Hilbert spaces and $\mathcal{H}$ their tensor product (defined in 1.1.3.). Further let T_1 and T_2 be operators with domains $D(T_1)$, $D(T_2)$ on $\mathcal{H}_1$ and $\mathcal{H}_2$ respectively. The <u>tensor product</u> $T = T_1 \otimes T_2$ of T_1 and T_2 is defined on $\mathcal{H}$ by setting $D(T)$ to be the pre-Hilbert space generated by vectors of the form $\psi = \psi_1 \otimes \psi_2$, with $\psi_1 \in D(T_1)$ and $\psi_2 \in D(T_2)$ and by $T\psi = T_1\psi_1 \otimes T_2\psi_2$.

<u>If T_1 and T_2 are essentially self-adjoint operators then it follows that the operators</u>

$$T_1 \otimes T_2 \quad , \quad T_1 \otimes \mathbb{1}_2 \; , \; \mathbb{1}_1 \otimes T_2 \quad , \quad T_1 \otimes \mathbb{1}_2 + \mathbb{1}_1 \otimes T_2$$

<u>are essentially self-adjoint on $\mathcal{H}$</u> . ($\mathbb{1}_1$ and $\mathbb{1}_2$ denote the identity operators on $\mathcal{H}_1$ and $\mathcal{H}_2$).

This last statement is again an immediate consequence of the criteria given in 1.1.10. Each of the operators under consideration is symmetric by definition but if ψ_1 and ψ_2 are analytic vectors for T_1 and T_2 then $\psi_1 \otimes \psi_2$ is an analytic vector for these operators. The density of the set of analytic vectors constructed in the manner is a consequence of the assumed essential self-adjointness of T_1 and T_2 .

<u>1.1.13.</u> We next illustrate some of the foregoing concepts by considering the operators of differentiation and double differentiation as operators on the Hilbert space $L^2(0, L)$ of complex functions over the open interval $(0, L)$ of the real line.

First let us introduce a number of domains suitable for the definition of differential operators. Define D_1 to be the set of all $\psi \in L^2(0, L)$ such that ψ is absolutely continuous on $[0, L]$ and has a derivative belonging to $L^2(0,L)$ Explicitly D_1 is the set of ψ expressible in the form

$$\psi(x) = c + \int_0^x dy\, \varphi(y)$$

where $c \in \mathbb{C}$ and $\varphi \in L^2(0,L)$; we identify c as $\psi(0)$ and φ as the derivative ψ' of ψ. Next define D_∞ and $D(\theta)$ by

$$D_\infty = \{ \psi \ ; \ \psi \in D_1 \ , \ \psi(0) = 0 = \psi(L) \}$$

$$D(\theta) = \{ \psi \ ; \ \psi \in D_1 \ , \ \psi(0) = e^{i\theta}\psi(L) \}$$

where $0 \leq \theta \leq 2\pi$ (the significance of the suffix ∞ will only become clear at a later stage).

Finally we define three operators, each corresponding to simple differentiation; P_0 , P_∞ and $P(\theta)$ by

$$D(P_0) = D_1 \quad \text{and} \quad (P_0\psi)(x) = i\,\psi'(x)$$

$$D(P_\infty) = D_\infty \quad \text{and} \quad (P_\infty\psi)(x) = i\,\psi'(x)$$

$$D(P(\theta)) = D(\theta) \quad \text{and} \quad (P(\theta)\psi)(x) = i\,\psi'(x) \ .$$

Although these operators all have the same action their properties differ greatly due to the different domains of definition $D(P_\infty) \subset D(P(\theta)) \subset D(P_0)$.

Lemma 1.1.14.

a. <u>P_∞ is a closed, symmetric, operator with adjoint $P_\infty^* = P_0$.</u>

b. <u>$P(\theta)$ is a self-adjoint extension of P_∞ ; the spectrum of $P(\theta)$ is discrete and consists of the eigenvalues $(2\pi n - \theta)/L$, $n = 0, \pm 1, \pm 2, \dots$ with corresponding eigenfunctions $\exp\{-ix(2\pi n - \theta)/L\}$.</u>

c. <u>P_0 is closed operator and $P_0^* = P_\infty$.</u>

A simple calculation involving partial integration demonstrates that P_∞ is symmetric. We will conclude that P_∞ and P_0 are closed by demonstrating that $P_\infty^* = P_0$, $P_0^* = P_\infty$ and then applying the last criterion of 1.1.6.

Let us calculate the adjoint of P_0 . Take $\varphi \in D(P_0^*)$ and set $\varphi^* = P_0^*\varphi$; then for each $\psi \in D(P_0)$ one has

$$(\varphi, P_0\psi) = (P_0^*\varphi, \psi) = (\varphi^*, \psi) = \int_0^L dx \left\{ \frac{d}{dx}\left[\int_0^x dy\, \overline{\varphi^*(y)} + \bar{c} \right] \right\} \psi(x)$$

where $c \in \mathbb{C}$ is arbitrary. Integrating by parts one finds

$$(\varphi, P_0\psi) = \left[\int_0^L dx\, \overline{\varphi^*(x)} + \bar{c}\right]\psi(L) - \bar{c}\,\psi(0) - \int_0^L dx\left[\int_0^x dy\, \overline{\varphi^*(y)} + \bar{c}\right]\frac{d\psi(x)}{dx}$$

and hence one must have

$$\int_0^L dx\left[i\bar{\varphi}(x) + \int_0^x dy\, \overline{\varphi^*(y)} + \bar{c}\right]\frac{d\psi(x)}{dx} = \left[\int_0^L dx\, \overline{\varphi^*(x)} + \bar{c}\right]\psi(L) - \bar{c}\,\psi(0) .$$

But first taking $\psi \in D(P_\infty)$, so that $\psi(0) = 0 = \psi(L)$, and noting that the range of P_∞ is dense in $L^2(0,L)$ we conclude that

$$\varphi(x) = -i\int_0^x dy\, \varphi^*(y) - ic .$$

Secondly noting that there are $\psi \in D(P_0)$ such that $\psi(0) \neq 0$ and $\psi(L) = 0$ and vice-versa, we also have

$$c = 0 \quad , \quad c + \int_0^L dx\, \varphi^*(x) = 0 . \qquad (*)$$

Thus we have deduced that

$$\varphi(x) = -i\int_0^x dy\, \varphi^*(y)$$

and $\varphi(0) = 0 = \varphi(L)$, i.e. $\varphi \in D(P_\infty)$. But we now have

$$(P_0^*\varphi)(x) = \varphi^*(x) = i\frac{d\varphi}{dx}(x)$$

and hence

$$P_0^* = P_\infty .$$

Similarly if we calculate the adjoint of P_∞ we find that each $\varphi \in D(P_\infty^*)$ must be in $D(P_0)$ but the boundary condition $(*)$ is absent. Thus $P_\infty^* = P_0$ and both P_0 and P_∞ are closed.

If we use the same method to calculate the adjoint of $P(\theta)$ the result is similar but the boundary condition $(*)$ is replaced by

$$ce^{i\theta} = \int_0^L dx\, \varphi^*(x) + c$$

i.e. $$\varphi(0) = e^{i\theta}\varphi(L) .$$

Thus $P^*(\theta) = P(\theta)$ and $P(\theta)$ is self-adjoint by definition.

The spectral properties of $P(\theta)$ are straightforwardly calculated.

1.1.15. Note that if we define restrictions of the operators P_∞, $P(\theta)$, P_0 by restricting their domains by conditions of differentiability then the argument of 1.1.14. establishes that the restricted operators are nevertheless closable and their closures coincide with P_∞, $P(\theta)$ and P_0.

Similarly we can define restrictions by specifying conditions on the boundary values of the first or higher derivatives without affecting the closure properties. For example if D is the set of functions which are absolutely continuous on the closed interval $[0,L]$ with $\psi' \in L^2(0,L)$ and $\psi'(0) = 0 = \psi'(L)$ then the closure of the restriction of P_0 to D is again P_0 , i.e. D is a core of P_0 (cf. 1.1.5.).

<u>1.1.16.</u> The definition of partial differential operators can be handled similarly but care has to be taken in defining the domains and the sense of differentiation. We will illustrate this problem and avoid the differentiability problems by defining closable as opposed to closed operators.

Let Λ be an open connected set of ν-dimensional Euclidean space and consider the Hilbert space $L^2(\Lambda)$. Let $C^1(\Lambda)$ be the set of once continuously differentiable functions over Λ and $C_0^1(\Lambda)$ the subset of $C^1(\Lambda)$ formed by the functions with the property $\psi(x) = 0$, $x \in \partial\Lambda$ ($\partial\Lambda$ denotes the surface of $\bar{\Lambda}$).

Now define the vector-valued operators $\underline{P}_0$, $\underline{P}_\infty$ by

$$D(\underline{P}_0) = C^1(\Lambda) \quad , \quad (\underline{P}_0 \psi)(x) = i \,\underline{\nabla}_x \psi(x) \quad , \quad \psi \in D(\underline{P}_0) \quad ,$$

$$D(\underline{P}_\infty) = C_0^1(\Lambda) \quad , \quad (\underline{P}_\infty \psi)(x) = i \,\underline{\nabla}_x \psi(x) \quad , \quad \psi \in D(\underline{P}_\infty) \quad .$$

In the special case that Λ is a parallelepiped,

$$\Lambda = \{ x = (x^{(1)}, \dots, x^{(\nu)}) \; ; \; 0 \le x^{(i)} < L^{(i)} \; , \; i = 1, 2, \dots \nu \}$$

we also define $\underline{P}(\theta)$ by

$$D(\underline{P}(\theta)) = \{ \psi \; ; \; \psi \in C^1(\Lambda), \; \psi(x^{(1)}, \dots, 0, \dots x^{(\nu)}) = e^{i\theta_i} \psi(x^{(1)}, \dots, L^{(i)}, \dots x^{(\nu)}), \; i = 1, 2, \dots \}$$

where $0 \le \theta_i < 2\pi$ and

$$(\underline{P}(\theta) \psi)(\underline{x}) = i \,\underline{\nabla}_x \psi(\underline{x}) \quad , \quad \psi \in D(P(\theta))$$

<u>Lemma</u> :

<u>$\underline{P}_\infty$, $\underline{P}(\theta)$ and $\underline{P}_0$ are closable operators.</u>

By partial integration one checks straightforwardly that $\underline{P}_\infty$ is symmetric and hence closable (1.1.7.).

Again using partial integration one finds that

$$(\underline{P}_\infty \varphi, \psi) = (\varphi, \underline{P}_0 \psi) \quad , \quad \varphi \in D(\underline{P}_\infty) \; , \; \psi \in D(\underline{P}_0)$$

i.e. $\underline{P}_\infty$ and $\underline{P}_0$ are adjoint to each other. Hence $\underline{P}_\infty^*$ is an extension of $\underline{P}_0$.

But P_∞ is densely defined and hence P_∞^* is closed (1.1.6.). Thus P_0 is closable (P_∞^* is a closed extension).

The closability of $P(\theta)$ follows by an identical argument.

Note that although the abstract reasoning of this paragraph is more straightforward than the calculations of 1.1.14. , the results are weaker, e.g. we have not established that $P_\infty^* = P_0^{**}$, $P_0^* = P_\infty^{**}$ or $P(\theta)^* = P(\theta)^{**}$.

1.1.17. Let us return to the consideration of the operation of double differentiation in one dimension, i.e. on the space $L^2(0,L)$ of 1.1.13.
We first define a minimal operator S by taking $D(S) = C_0^2(0,L)$ the twice continuously differentiable functions over $(0,L)$, which vanish with their first derivatives at the end points 0 and L , and then set

$$(S\psi)(x) = -\frac{d^2\psi(x)}{dx^2} \quad , \quad \psi \in D(S) .$$

Lemma :

The operator S is symmetric. The three operators

$$S_\infty = P_0 P_\infty = P_\infty^* P_\infty$$

$$S(\theta) = [P(\theta)]^2 = P(\theta)^* P(\theta)$$

$$S_0 = P_\infty P_0 = P_0^* P_0$$

are self-adjoint extensions of S . The domains of these extensions are the subsets of the absolutely continuous functions ψ over $(0,L)$ with absolutely continuous first derivatives ψ' such that $\psi'' \in L^2(0,L)$, specified by the three boundary conditions

$$B_\infty \; ; \quad \psi(0) = 0 = \psi(L)$$

$$B(\theta) \; ; \quad \psi(0) = e^{i\theta}\psi(L) \quad , \quad \psi'(0) = e^{i\theta}\psi'(L)$$

$$B_0 \; ; \quad \psi'(0) = 0 = \psi'(L)$$

respectively.

The symmetry of S is again straightforwardly checked by partial integration.

We have established in 1.1.14. that $P_0 = P_\infty^*$, $P(\theta) = P(\theta)^*$ and $P_\infty = P_0^*$ and hence the alternative forms of S_∞ , $S(\theta)$ and S_∞ are valid.

But P_∞ , $P(\theta)$ and P_0 are densely defined and closed (1.1.14.) hence S_∞ , $S(\theta)$ and S_0 are self-adjoint by the criterion of 1.1.11.

The statement concerning domains is a consequence of the representations $S_\infty = P_0 P_\infty$ etc. and the definition of the product of operators (1.1.4.).

1.1.18. The self-adjoint extensions of S described in 1.1.17. do not exhaust all possibilities. All extensions can however be characterized by defining operators of double differentiation on the domain mentioned in the foregoing lemma and then taking restrictions defined by boundary conditions of a more general form than those S_∞ etc. In fact a complete characterization of all self-adjoint extensions of S is given by the family of boundary conditions

$$\begin{pmatrix} \alpha_{11} & \alpha_{12} \\ \alpha_{21} & \alpha_{22} \end{pmatrix}\begin{pmatrix} \psi(0) \\ \psi'(0) \end{pmatrix} + \begin{pmatrix} \beta_{11} & \beta_{12} \\ \beta_{21} & \beta_{22} \end{pmatrix}\begin{pmatrix} \psi(L) \\ \psi'(L) \end{pmatrix} = \begin{pmatrix} 0 \\ 0 \end{pmatrix}$$

where the complex matrix coefficients must satisfy the conditions

$$\overline{\alpha}_{11}\alpha_{12} - \alpha_{11}\overline{\alpha}_{12} = \overline{\beta}_{11}\beta_{12} - \beta_{11}\overline{\beta}_{12}$$

$$\overline{\alpha}_{21}\alpha_{22} - \alpha_{21}\overline{\alpha}_{22} = \overline{\beta}_{21}\beta_{22} - \beta_{21}\overline{\beta}_{22}$$

$$\overline{\alpha}_{11}\alpha_{22} - \alpha_{21}\overline{\alpha}_{12} = \overline{\beta}_{11}\beta_{22} - \beta_{21}\overline{\beta}_{12} .$$

We will not discuss all these extensions but restrict our attention to those obtained in 1.1.17. and a second class discussed below in 1.1.20.

We note in passing that although S is real there exists a continuum of self-adjoint extensions of S that are not real, e.g. $S(\theta)$ for $0 \leq \theta < 2\pi$.

In this context a self-adjoint extension T is real whenever $\psi \in D(T)$ and $T\psi = \varphi$ implies that $\overline{\psi} \in D(T)$ and $T\overline{\psi} = \overline{\varphi}$; the non-real extensions T of S arise because there are some $\psi \in D(T)$ such that $\overline{\psi} \notin D(T)$.

1.1.19. In the sequel we will need, for estimation purposes, spectral properties of the self-adjoint operators S_∞ , $S(\theta)$, S_0 introduced in 1.1.17. and a family of additional extensions to be introduced below. It can be established that each of these extensions has a discrete spectrum and we now list the set of eigenvalues of each operator and the associated unnormalized eigenfunctions. We also comment on the quantum mechanical interpretation of these operators as Hamiltonians of a particle constrained to move on the interval $(0, L)$.

$$S_\infty \quad \begin{array}{lll} \text{Eigenvalues} & (n\pi/L)^2 & n = 1, 2, \cdots \\ \text{Eigenfunctions} & \sin n\pi x/L & n = 1, 2, \cdots \end{array}$$

S_∞ is the Hamiltonian which corresponds to the motion of a particle constrained to move on the interval $(0, L)$ with a "repulsion" of the particle at each boundary 0 and L. The fact that each function in the domain of S_∞ must vanish at the boundaries indicates that the probability of finding a particle in the neighbourhood of the boundary points is small.

$$S(\theta) \begin{cases} \text{Eigenvalues} & (\theta + 2\pi n/L)^2 \qquad n = 0, \pm 1, \pm 2, \cdots \\ \text{Eigenfunctions} & \exp\{ix(\theta + 2\pi n)/L\} \qquad n = 0, \pm 1, \pm 2, \cdots \end{cases}$$

The $S(\theta)$ are Hamiltonians which correspond to periodic motion of the particle on the interval $(0, L)$. The particle moves freely within the interval but on arriving at a boundary is transmitted and re-enters at the opposite boundary (with its wave amplitude changed by a phase). Note that $S(\theta)$ has the "free Hamiltonian" representation as a momentum squared $S(\theta) = P(\theta)^2$ (cf. 1.1.17.).

1.1.20 Instead of discussing the spectrum of S_0 directly we consider it as a special case of the class of operators defined as follows. S_σ is an operator depending upon two real parameters $\sigma = (\sigma_0, \sigma_L)$, $\sigma \in R^2$. The domain $D(S_\sigma)$ is the set of absolutely continuous functions ψ over $[0, L]$ with absolutely continuous first derivatives ψ', such that $\psi'' \in L^2(0, L)$, which satisfy the boundary conditions

$$B_\sigma : \qquad \psi'(0) = \sigma_0 \psi(0) \quad , \quad \psi'(L) = -\sigma_L \psi(L)$$

The action of S_σ is defined by

$$(S_\sigma \psi)(x) = -\frac{d^2\psi}{dx^2}(x) \quad , \quad \psi \in D(S_\sigma)$$

Note that for $\sigma = 0$, i.e. $\sigma_0 = 0 = \sigma_L$, this operator coincides with the S_0 of 1.1.17.

Lemma :

S_σ is a self-adjoint extension of S.

It is readily established by partial integration that S_σ is a symmetric extension of S. To prove that it is self-adjoint it suffices to establish that $S_\sigma^* = S_\sigma$. But the adjoint can be calculated in the same manner that we calculated P_∞^* in 1.1.14. Take $\varphi \in D(S_\sigma^*)$ and set $\varphi^* = S_\sigma^* \varphi$ then for each $\psi \in D(S_\sigma)$ one has

$$(\varphi^*, \psi) = (\varphi, S_\sigma \psi) .$$

Now one follows the calculation of 1.1.14., partially integrating and using the freedom of choice of ψ to conclude that $\varphi \in D(S_\sigma)$ and $\varphi^* = S_\sigma \varphi$. We leave the details of the calculation to the reader.

Let us now discuss the spectrum of S_σ . We will compute implicitly an infinite set of eigenfunctions but we will not prove that the spectrum of S_σ is purely discrete and that we have a description of all eigenvalues. (Actually this can be inferred from the mini-max theorem given in 1.2.14.) Consider the functions ψ_λ^σ defined by

$$\psi_\lambda^\sigma(x) = \cos\sqrt{\lambda}\, x + \frac{\sigma_0}{\sqrt{\lambda}} \sin\sqrt{\lambda}\, x \qquad , \qquad \lambda \in \mathbb{C}$$

These functions satisfy the boundary condition

$$\left[\frac{d\psi_\lambda^\sigma}{dx}(x) - \sigma_0 \psi_\lambda^\sigma(x) \right]_{x=0} = 0 \qquad ,$$

and satisfy

$$\left[\frac{d\psi_\lambda^\sigma}{dx}(x) + \sigma_L \psi_\lambda^\sigma(x) \right]_{x=L} = 0$$

if and only if

$$\frac{1}{\sqrt{\lambda}} \tan\sqrt{\lambda}\, L = \frac{(\sigma_0 + \sigma_L)}{\lambda - \sigma_0\sigma_L} \qquad (*) .$$

Let λ_n be a root of this equation we can then conclude that $\psi_{\lambda_n}^\sigma \in D(S_\sigma)$ and

$$S_\sigma \psi_{\lambda_n}^\sigma = \lambda_n \psi_{\lambda_n}^\sigma .$$

It is easily established that there are an infinite number of roots of $(*)$ and hence we have an implicit description of the eigenvalues and eigenfunctions of S_σ.

Let us consider the special case $\sigma_0 = \sigma_L = \sigma$ in more detail. First two general remarks. As S_σ is self-adjoint the eigenvalues λ_n must be real and hence it suffices to find real or pure imaginary values of $\sqrt{\lambda}$ satisfying $(*)$. Secondly for each root $\sqrt{\lambda_n}$ of $(*)$ there is a second root $-\sqrt{\lambda_n}$ which leads to the identical eigenvalue and eigenfunction thus we need only consider $\sqrt{\lambda} \geq 0$ or $-i\sqrt{\lambda} > 0$.

We will consider the three cases $\sigma > 0$, $\sigma = 0$ and $\sigma < 0$ separately and solve $(*)$ by graphical means.

Case 1 : $\sigma > 0$ Set $f(\mu) = \mu^{-1} \tan \mu L$, $g(\mu) = 2\sigma(\mu^2 - \sigma^2)^{-1}$;

the graphs of these functions are shown in fig. 1 .

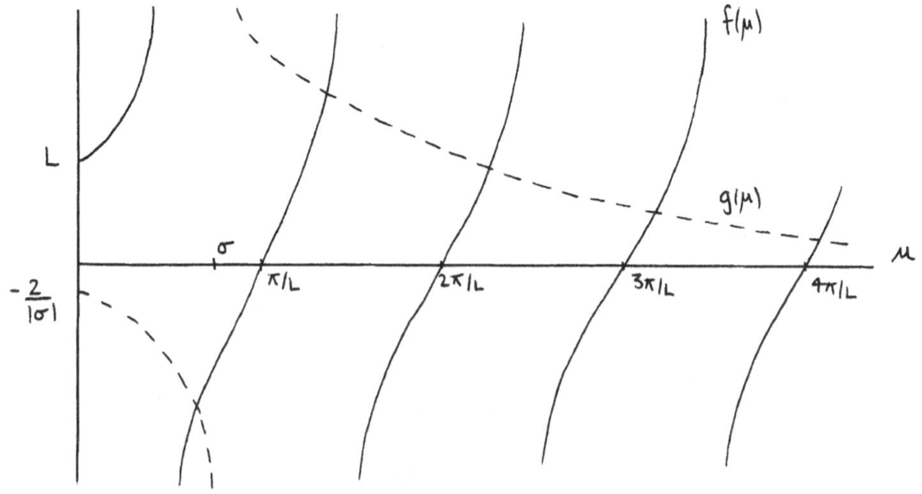

Fig. 1

One deduces that the eigenvalues $(\lambda_n(\sigma))_{n\geq 1}$ of S_σ form an increasing sequence with the property

$$0 < \lambda_1(\sigma) < (\pi/L)^2 < \lambda_2(\sigma) < (2\pi/L)^2 < \cdots\cdots$$

(it is straightforward to check that for $\sigma > 0$ there are no purely imaginary roots of $(*)$).

Note that each $\lambda_n(\sigma)$ is a monotonically increasing function of σ with

$$\lim_{\sigma\to\infty} \lambda_n(\sigma) = (n\pi/L)^2 \quad , \quad \lim_{\sigma\to 0} \lambda_n(\sigma) = ((n-1)\pi/L)^2 \ .$$

It also follows that the eigenvalues are monotonically decreasing functions of L .

Case 2 : $\sigma = 0$ The eigenvalue equation $(*)$ is directly soluble in this case and the eigenvalues $\{\lambda_n(0)\}_{n\geq 1}$ of S_0 are given in increasing order by :

$$\lambda_n(0) = ((n-1)\pi/L)^2 \qquad n = 1, 2, \cdots\cdots .$$

Note that for each $n\geq 1$ one has :

$$\lambda_n(0) = \lim_{\sigma\to 0} \lambda_n(\sigma)$$

Case 3 : $\sigma < 0$. This latter case is more complicated because there are purely imaginary roots of $(*)$. Setting $\sqrt{\lambda} = i\mu$ we are led to the examination of the equations $f(\mu) = g(\mu)$ with

$$f(\mu) = \tanh \mu L \qquad g(\mu) = 2|\sigma|\mu/(\mu^2+\sigma^2) \qquad \mu > 0 .$$

This is illustrated in figures 2 and 3 for the cases $2 < |\sigma|L$ and $2 > |\sigma|L$.

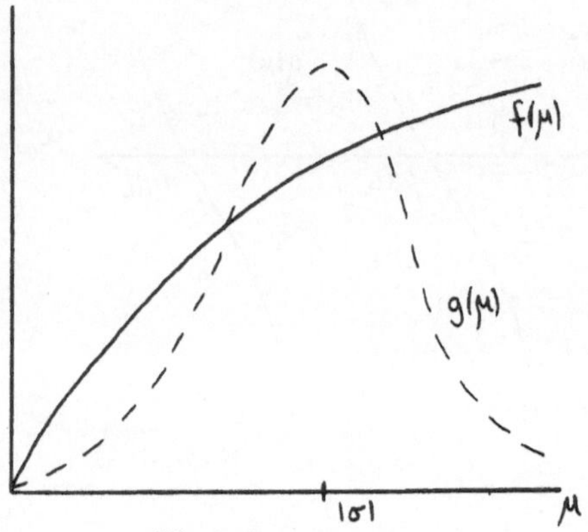

Fig. 2

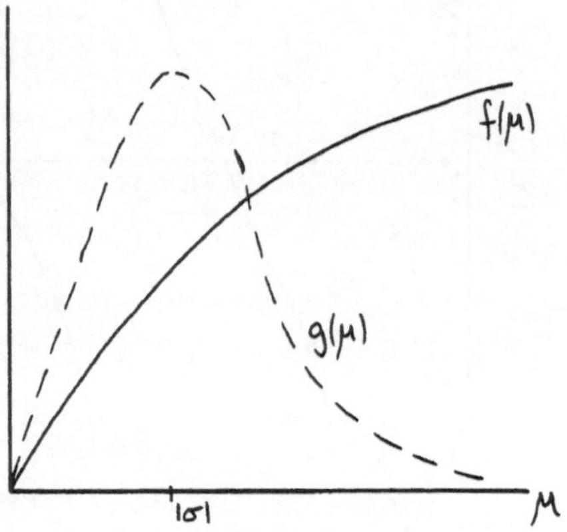

Fig. 3

Thus in the first case there are two non-zero roots and in the second case one.

Now consider the equation $(*)$ with $\sqrt{\lambda} = \mu \geqslant 0$. We compare the graphs of

$$f(\mu) = \mu^{-1} \operatorname{Tan} \mu L \qquad \text{and} \qquad g(\mu) = -2|\sigma|/(\mu^2 - |\sigma|^2)$$

in figures 4 and 5 for the two cases $2 < |\sigma|L$ and $2 > |\sigma|L$.

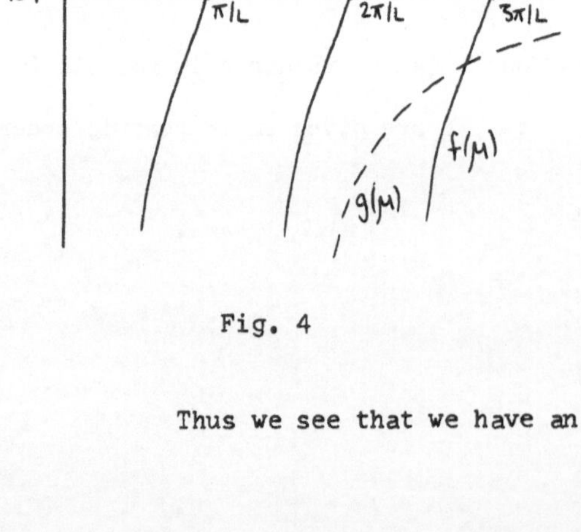

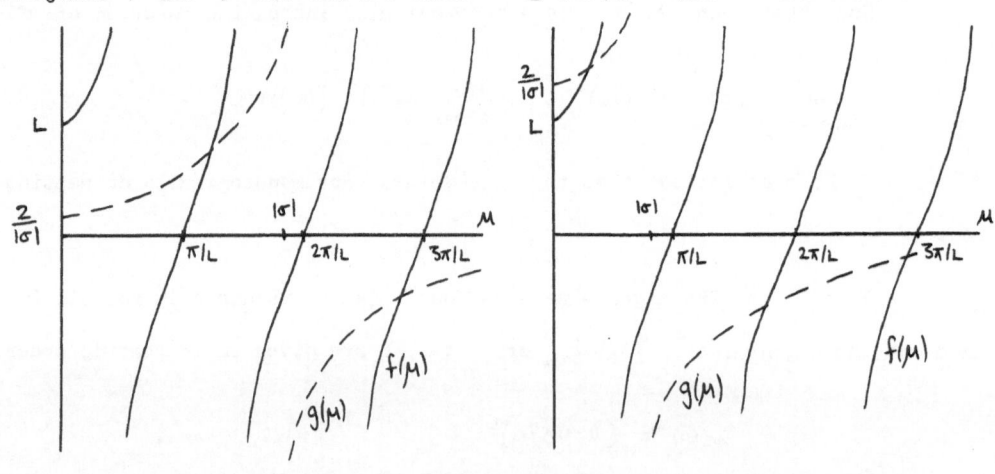

Fig. 4

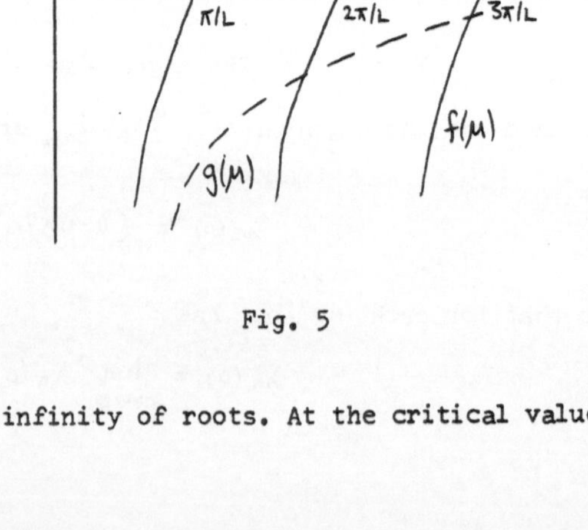

Fig. 5

Thus we see that we have an infinity of roots. At the critical value

$L|\sigma| = 2$ one purely imaginary root depicted in figure 2 passes into a real root appearing in figure 5 . Thus if $\{\lambda_n(\sigma)\}_{n\geq 1}$ are the eigenvalues we have the two cases

$$|\sigma| > 2/L \quad ; \quad \lambda_1(\sigma) < \lambda_2(\sigma) < 0 \quad , \quad (\pi/L)^2 < \lambda_3(\sigma) < (2\pi/L)^2 < \cdots$$

$$|\sigma| < 2/L \quad ; \quad \lambda_1(\sigma) < 0 < \lambda_2(\sigma) < (\pi/L)^2 < \lambda_3(\sigma) < (2\pi/L)^2 < \cdots$$

and further

$$\lambda_n(0) = ((n-1)\pi/L)^2 = \lim_{\sigma \to 0-} \lambda_n(\sigma) \; .$$

Note that for $\sigma < 0$ fixed the behaviour of the $\lambda_n(\sigma)$ as functions of L is such that $\lambda_1(\sigma)$ is monotonically increasing to the asymptotic value and the , are monotonically decreasing ; and , (cf. figure 6).

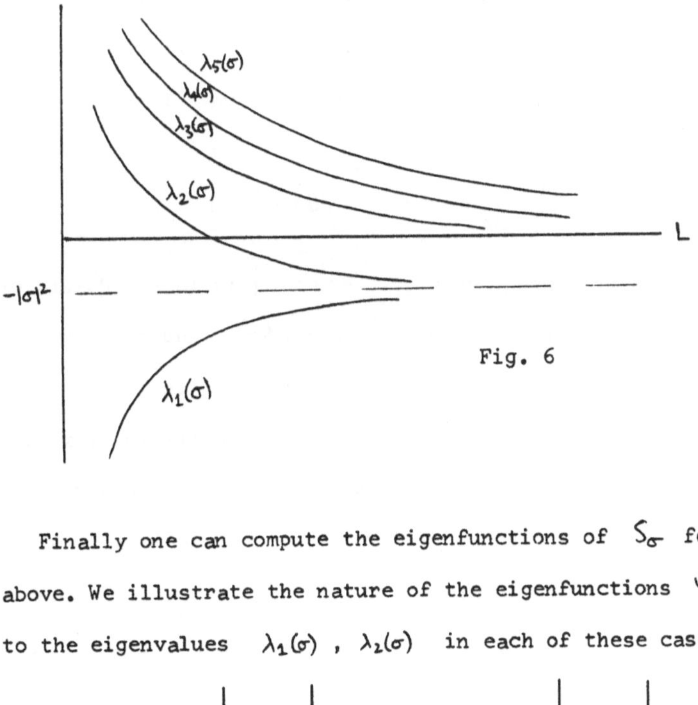

Fig. 6

Finally one can compute the eigenfunctions of S_σ for the three cases considered above. We illustrate the nature of the eigenfunctions $\psi_1(\sigma)$, $\psi_2(\sigma)$ corresponding to the eigenvalues $\lambda_1(\sigma)$, $\lambda_2(\sigma)$ in each of these cases in figure 7 .

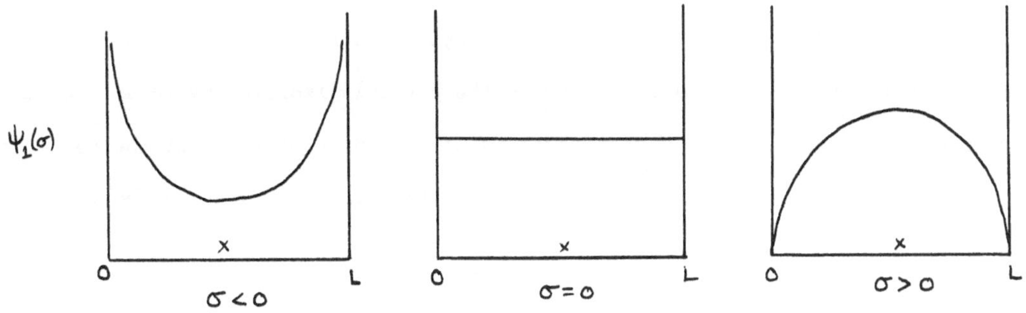

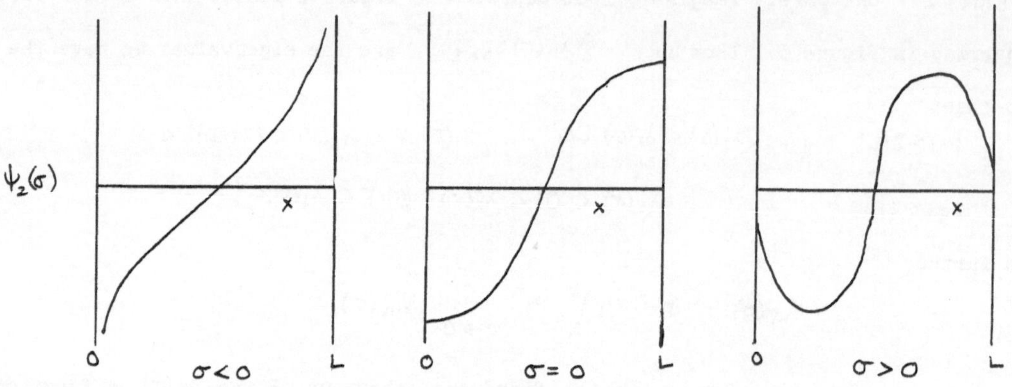

Fig. 7

Let us now consider the quantum mechanical interpretation of S_σ and S_∞ as the Hamiltonians of a particle moving on the interval $(0,L)$ in more detail For each $\psi \in L^2(0,L)$ the function $x \to |\psi(x)|^2$ represents the probability density distribution of the particle described by the wave function ψ and the function $x \to j(x)$ where

$$j(x) = i\left[\frac{d\bar{\psi}}{dx}(x)\,\psi(x) - \bar{\psi}(x)\frac{d\psi}{dx}(x)\right]$$

represents the probability current density. Now if ψ is in the domain of one of the operators S_σ or S_∞ then $j(0) = 0 = j(L)$ and it is easily checked from 1.1.18. that the latter conditions characterize this set of self-adjoint extensions of S. Thus for any one of these operators the particle flux both into and out of the system is zero, i.e. the operators correspond to the Hamiltonians of truly isolated systems. On the other hand if ψ is in the domain of one of the other possible extensions of S, for example $S(\theta)$, one has only the weaker condition $j(0) = j(L)$;what flows in at one end flows out of the other.

Now let us briefly explain the significance of the parameter σ which characterizes the above operators and hence the possible isolated systems. We see from figure 7 that if $\sigma < 0$ then there is a large probability that the particle in its ground state is near the boundaries, i.e. the particle is attracted to the boundaries.

Conversely if $\sigma > 0$ the particle is repelled from the boundaries. Thus σ is a measure of the attractiveness, or elasticity of the boundaries, the case $\sigma = 0$ corresponding to the case of perfect elasticity and the cases $\sigma > 0$, $\sigma < 0$ corresponding to repulsion and attraction respectively.

1.1.21 To conclude our discussion of this class of differential operators note that S_∞ is, in a sense which we have not made precise, the limit of S_σ as $\sigma \to \infty$ In fact if $\{\lambda_n(\infty)\}_{n\geq 1}$ denote the eigenvalues of S_∞ in increasing order we have at least seen that

$$\lambda_n(\infty) = \lim_{\sigma \to +\infty} \lambda_n(\sigma) .$$

The behaviour of the $\lambda_n(\sigma)$ for $\sigma \to -\infty$ is much different for $n = 1$ and $n = 2$. We illustrate the behaviour of the set of eigenvalues as a function of σ^{-1} in fig. 8.

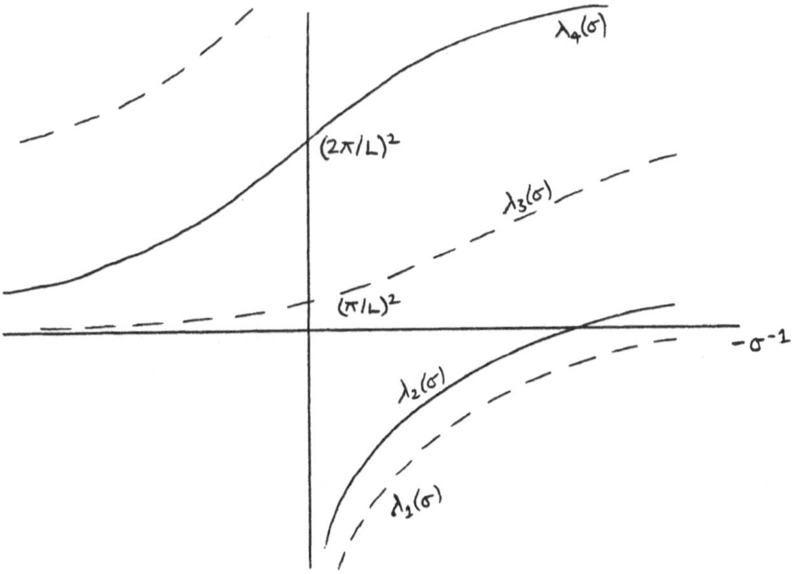

Fig. 8

Note that the eigenvalues can be considered as given by two functions in the manner indicated in figure 8 .

§2 - POSITIVE FORMS AND SELF-ADJOINT OPERATORS

1.2.1. The theory of positive forms and the theory of positive self-adjoint operators are closely related. We develop next a few of the concepts of the former theory with the aim of obtaining a suitable method of dealing with properties of the operators occurring as Hamiltonians of finite systems. It will be seen that many of of the definitions introduced below are analogous to those given in §1 for operators Throughout this section we deal with a complex Hilbert space $\mathcal{H}$.

1.2.2. A sesquilinear form t is a map from the Cartesian product $D(t)\times D(t)$ of a set $D(t)\subset\mathcal{H}$, the domain of t , to the complex numbers $\mathbb{C}$ which is anti-linear in the first component and linear in the second. We use the notation

$$\varphi,\psi\in D(t) \xrightarrow{t} t(\varphi,\psi)\in\mathbb{C} .$$

Thus if $\psi_1,\psi_2,\varphi\in D(t)$ and $\alpha_1,\alpha_2\in\mathbb{C}$ we have $\alpha_1\psi_1+\alpha_2\psi_2\in D(t)$ and

$$t(\alpha_1\psi_1+\alpha_2\psi_2,\varphi) = \bar{\alpha}_1 t(\psi_1,\varphi)+\bar{\alpha}_2 t(\psi_2,\varphi)$$

$$t(\varphi,\alpha_1\psi_1+\alpha_2\psi_2) = \alpha_1 t(\varphi,\psi_1)+\alpha_2 t(\varphi,\psi_2) .$$

The form t is said to be densely defined if $D(t)$ is a dense subset of $\mathcal{H}$.

Each sesquilinear form determines uniquely a quadratic form defined by

$$t(\psi) = t(\psi,\psi) \qquad \psi\in D(t) .$$

Conversely $\{t(\psi)\,;\ \psi\in D(t)\}$ determines the sesquilinear form by the polarization formula

$$t(\varphi,\psi) = \frac{1}{4}\{t(\varphi+\psi)-t(\varphi-\psi)+it(\varphi+i\psi)-it(\varphi-i\psi)\} \qquad \varphi,\psi\in D(t) .$$

Two forms t_1 and t_2 are equal, $t_1=t_2$, if and only if they have the same domain D and $t_1(\varphi) = t_2(\varphi)$ for all $\varphi\in D$. Further t_1 is an extension of t_2 , if and only if $D(t_1)\supset D(t_2)$ and $t_1(\varphi)=t_2(\varphi)$ for all $\varphi\in D(t_2)$. The sum t_1+t_2 of the forms t_1 and t_2 is defined by

$$D(t) = D(t_1+t_2) = D(t_1)\cap D(t_2)$$

and

$$t(\psi) = t_1(\psi)+t_2(\psi) \qquad , \qquad \psi\in D(t)$$

and the product αt of t by $\alpha\in\mathbb{C}$ is given by

$$(\alpha t)(\psi) = \alpha t(\psi) \qquad , \qquad \psi\in D(\alpha t) = D(t) .$$

1.2.3. A sesquilinear form t is said to be symmetric if

$$t(\varphi,\psi) = \overline{t(\psi,\varphi)}$$

and from the polarization formula we see that t is symmetric if and only if the associated quadratic form is real-valued.

1.2.4. A symmetric form t is said to be bounded from below or lower semi-bounded if

$$t(\psi) \geq \gamma \|\psi\|^2 \qquad , \quad \psi \in D(t) .$$

The largest number γ with this property is called the lower bound of t and we write $t \geq \gamma$. In particular if $t \geq 0$ then t is said to be positive (more precisely, non-negative).

More generally an order relation is introduced between symmetric forms by defining

$t_1 \geq t_2$ if $D(t_1) \subset D(t_2)$ and

$$t_1(\psi) \geq t_2(\psi) \qquad , \quad \psi \in D(t_1) .$$

Note that this definition is slightly peculiar insofar the larger form has the smaller domain. Thus for example if $t_2 \supset t_1$ then $t_1 \geq t_2$. (The significance of this order will become somewhat clearer in 1.2.14.).

Each positive symmetric form t satisfies the inequalities

$$|t(\varphi,\psi)| \leq t(\varphi)^{\frac{1}{2}}\, t(\psi)^{\frac{1}{2}}$$

$$t(\varphi+\psi)^{\frac{1}{2}} \leq t(\varphi)^{\frac{1}{2}} + t(\psi)^{\frac{1}{2}}$$

$$t(\varphi+\psi) \leq 2t(\varphi) + 2t(\psi) .$$

1.2.5. Let t be a positive symmetric form. A sequence $\{\psi_n\}_{n\geq 1}$ of vectors in $D(t)$ is said to be t-convergent (to $\psi \in \mathcal{H}$) if for each $\epsilon > 0$ there is an n_ϵ such that

$$\|\psi_n - \psi_m\| < \epsilon$$

and

$$t(\psi_n - \psi_m) < \epsilon$$

for all $n, m > n_\epsilon$ (and ψ_n converges strongly to ψ).

The form t is defined to be closed if the t-convergence of $\{\psi_n\}_{n\geq 1}$ to ψ implies both that $\psi \in D(t)$ and

$$\lim_{n\to\infty} t(\psi_n - \psi) = 0 .$$

The form t is said to be closable if it has a closed extension. In particular t is closable if and only if the condition ψ_n is t-convergent to zero implies that $t(\psi_n) \to 0$. If this latter condition is satisfied then the closure (smallest closed extension) $\tilde{t}$ of t can be defined as follows. The domain $D(\tilde{t})$ is the set of all $\psi \in \mathcal{H}$ such that there exists a sequence $\{\psi_n\}_{n\geq 1}$ with ψ_n t-convergent to ψ and $\tilde{t}$ is given

$$\tilde{t}(\psi) = \lim_{n\to\infty} t(\psi_n) .$$

If t is closed a linear submanifold D' of $D(t)$ is called a <u>core</u> of t if the restriction t' of t with domain D' has the closure t, i.e. if $\tilde{t}' = t$.

<u>Proposition 1.2.6.</u> <u>Let t be a densely defined, closed, positive form on $\mathcal{H}$. There exists a positive self-adjoint operator T with domain $D(T)$ dense in $\mathcal{H}$ and such that</u>

1. $D(T) \subset D(t)$ and $t(\varphi, \psi) = (\varphi, T\psi)$

<u>for every $\varphi \in D(t)$ and $\psi \in D(T)$. The operator T is uniquely determined by this condition.</u>

2. <u>$D(T)$ is a core of t</u> .

3. <u>If $\psi \in D(T)$, $\chi \in \mathcal{H}$ and $t(\varphi,\psi) = (\varphi, \chi)$ for every φ in a core of t then $\psi \in D(T)$ and $T\psi = \chi$</u> .

4. $D(t) = D(T^{\frac{1}{2}})$ and $t(\varphi, \psi) = (T^{\frac{1}{2}}\varphi, T^{\frac{1}{2}}\psi)$, $\varphi,\psi \in D(t)$.

<u>$D \subset D(t)$ is a core of t if and only if it is a core of $T^{\frac{1}{2}}$</u> .

This proposition shows that the forms provide a convenient means of constructing and characterizing positive self-adjoint operators. For example in the construction of self-adjoint extensions of a given positive symmetric operator it is often easy to use the operator to construct symmetric forms. If these forms are closable etc. then they yield, by the above construction, the desired self-adjoint extensions. The only feature which is rather delicate is the closability but this can often be inferred from the following criteria.

<u>1.2.7.</u> Let S be an arbitrary operator on $\mathcal{H}$ and define t by $D(t) = D(S)$ and

$$t(\psi) = (S\psi, S\psi) \qquad \psi \in D(S)$$

then t is positive, symmetric. Further t is closable if and only if S is closable and t is closed if and only if S is closed.

This result follows immediately from the respective definitions of closability for operators and forms.

1.2.8. Let S be a positive symmetric operator on $\mathcal{H}$ and define t by $D(t) = D(S)$ and

$$t(\varphi, \psi) = (\varphi, S\psi) \qquad \varphi, \psi \in D(S)$$

then t is positive, symmetric, and closable.

This result is slightly more difficult to deduce. It was the original result, due to Friederichs, which led to the development of the theory of positive forms in association with positive self-adjoint operators ; for this reason the self-adjoint operator associated with the closure of t is often referred to as the Friederichs extension of S.

1.2.9. One can also construct closed forms by addition or limit processes. For example if t_1 and t_2 are two positive closed (closable) forms then the sum $t_1 + t_2$ is also positive and closed (closable). Of course the sum is not necessarily densely defined. Similarly if $\{t_n\}_{n\geq 1}$ is a family of positive closed (closable) forms and their sum t is defined by

$$t(\psi) = \sum_{n\geq 1} t_n(\psi)$$

with domain $D(t)$ given as the $\psi \in \bigcap_n D(t_n)$ such that the sum converges then t is closed (closable). (The proof of this statement is essentially identical to the proof that the direct sum Hilbert space defined in 1.1.2. is complete).

The last result can be rephrased in the following manner. Let be a monotonically increasing family of positive closed (closable) forms i.e. $t_1 \leq t_2 \leq t_3 \dots$ then we can define the limit form t by

$$t(\psi) = \lim_{n\to\infty} t_n(\psi) = \sup_n t_n(\psi)$$

with the domain $D(t)$ now specified to be the $\psi \in \bigcap_n D(t_n)$ such that the supremum is finite, and it follows that t is closed (closable). Again t is not necessarily densely defined.

1.2.10. Let $\{t_n\}_{n\geq 1}$ be a monotonically increasing family of positive closed forms, let t be their limit defined in 1.2.9. and assume t is densely defined. Let $\{T_n\}_{n\geq 1}$ and T denote the self-adjoint operators associated with the foregoing forms by 1.2.6. and assume that these operators have discrete spectrum with finite multiplicity. Denote the eigenvalues, arranged in increasing order and repeated according to multiplicity, of T_n by $\{\lambda_m(T_n)\}_{n\geq 1}$ etc. It follows that

$$\lim_{n\to\infty} \lambda_m(T_n) = \lambda_m(T) \quad .$$

Again this result does not seem to appear in the standard literature so we will indicate a proof which uses the minimax theorem of 1.2.14.

For simplicity we only consider the lowest eigenvalue and leave the similar discussion of the higher eigenvalues to the reader. We adopt the notation $\lambda_n = \lambda_1(T_n)$, $\lambda = \lambda_1(T)$. From 1.2.14. we have $\lambda_n \leq \lambda_{n+p} \leq \lambda$, $p \geq 0$ and hence introducing $\bar{\lambda} = \sup_n \lambda_n$ we have $\lambda \geq \bar{\lambda}$.

Next let $\{\psi_n\}$ denote a set of normalized eigenfunctions of $\{T_n\}$ corresponding to the eigenvalues $\{\lambda_n\}$. As the $\{\psi_n\}$ are normalized we can assume, possibly passing to a subsequence, that ψ_n converges weakly to a vector ψ . But if $n < m, p$ and $E_n(\mu)$ denotes the projector on the subspace spanned by the eigenvectors of T_n with eigenvalues less than μ then

$$\begin{aligned} 4\bar{\lambda} &\geq 2t_m(\psi_m) + 2t_p(\psi_p) \\ &\geq 2t_n(\psi_m) + 2t_n(\psi_p) \\ &\geq t_n(\psi_m - \psi_p) \\ &\geq t_n((1-E_n(\mu))(\psi_m - \psi_p)) \\ &\geq \mu \, \|(1-E_n(\mu))(\psi_m - \psi_p)\|^2 \,. \end{aligned}$$

Thus

$$\|\psi_m - \psi_p\|^2 \leq \|E_n(\mu)(\psi_m - \psi_p)\|^2 + 4\bar{\lambda}/\mu \quad .$$

Now given $\epsilon > 0$ we can choose μ such that $8\bar{\lambda}/\mu < \epsilon$ and then because $E_n(\mu)$ is finite dimensional and the ψ_n converge weakly, we can choose $n_\epsilon(\mu)$ such that $\|E_n(\mu)(\psi_m - \psi_p)\|^2 < \epsilon/2$ for $m, p > n_\epsilon(\mu)$. Thus

$$\|\psi_m - \psi_p\|^2 < \epsilon \qquad m, p > n_\epsilon(\mu)$$

i.e. ψ_n converges strongly.

Next we have

$$\|\psi_m + \psi_p\|^2 = 2\|\psi_m\|^2 + 2\|\psi_p\|^2 - \|\psi_m - \psi_p\|^2 \geq 4 - \epsilon \qquad m, p > n_\epsilon(\mu)$$

and for $n < m < p$

$$\begin{aligned} t_n(\psi_m - \psi_p) &\leq t_m(\psi_m - \psi_p) \\ &= 2t_m(\psi_m) + 2t_m(\psi_p) - t_m(\psi_m + \psi_p) \\ &\leq 2(\lambda_m + \lambda_p) - t_m(\psi_m + \psi_p) \\ &\leq 2(\lambda_m + \lambda_p) - \lambda_m \|\psi_m + \psi_p\|^2 \\ &\leq 2(\lambda_p - \lambda_m) + \epsilon\bar{\lambda} \end{aligned}$$

where the fourth step uses 1.2.14. But this estimate shows that ψ_n is t_n-convergent.

Finally from 1.2.14. and this convergence property we have

$$\lambda \leq \sup_n t_n(\psi) = \sup_n \lim_{m\to\infty} t_n(\psi_m) .$$

But the monotonicity of the forms gives

$$\lim_{m\to\infty} t_n(\psi_m) \leq \lim_{m\to\infty} t_m(\psi_m) = \bar{\lambda} .$$

Thus we have $\bar{\lambda} \leq \lambda \leq \bar{\lambda}$ and the result is established.

<u>1.2.11.</u> Let us consider the addition of forms in more detail. It is natural to ask what conditions two forms t_1 and t_2 must satisfy to ensure that $t_1 + t_2$ is bounded below. Clearly this is the case if t_1 and t_2 are bounded below but this condition is not necessary. One weaker condition can be found by considering the concept of relative boundedness. Let t_1 be bounded below then t_2 is said to be <u>t_1-bounded from below</u> with bound b if $D(t_2) \supset D(t_1)$ and

$$t_2(\psi) \geq -a\|\psi\|^2 - b\,t_1(\psi) \qquad , \quad \psi \in D(t_1)$$

with $a, b \geq 0$. If t_2 is t_1-bounded from below with bound $b \leq 1$ then

$$(t_1 + t_2)(\psi) \geq -a\|\psi\|^2 + (1-b)\,t_1(\psi) \qquad , \quad \psi \in D(t_1)$$

i.e. $t_1 + t_2$ is bounded from below.

Further t_2 is said to be <u>t_1-bounded</u> with bound b if $D(t_2) \supset D(t_1)$ and

$$|t_2(\psi)| \leq a\|\psi\|^2 + b\,|t_1(\psi)| \qquad , \quad \psi \in D(t_1)$$

with $a, b \geq 0$. If t_2 is t_1-bounded with bound $b < 1$ then $t_1 + t_2$ is bounded from below and $t_1 + t_2$ is closed (closable) if and only if t_1 is closed. (closable).

1.2.12. If t_1 and t_2 are positive closed forms and their sum $t = t_1 + t_2$ is densely defined then there are self-adjoint operators T_1, T_2 and T associated with the three forms and T may be regarded as the generalized sum of T_1 and T_2 which we denote by

$$T = T_1 \dot{+} T_2 .$$

Conversely if T_1 and T_2 are self-adjoint operators which are bounded below the associated forms exist and the generalized sum can be defined as the operator associated with $t = t_1 + t_2$ whenever this latter form is densely defined.

The generalized sum is an extension of the ordinary sum and in general the two do not coincide. If however $T_1 + T_2$ is essentially self-adjoint then $T_1 \dot{+} T_2 = (T_1 + T_2)^{**}$ from part 3 of 1.2.6.

1.2.13. Although in the foregoing we have principally spoken of positive forms one can easily deal with lower semi-bounded forms.

For example if $t \geq \gamma$ then t' defined by

$$t'(\psi) = t(\psi) - \gamma \|\psi\|^2 \qquad , \quad \psi \in D(t) = D(t')$$

is positive. If t is densely defined and closed the same is true of t' and we may associate the operator T' to t' and then the operator $T = T' + \gamma 1$ has as consequence the property

$$t(\varphi, \psi) = (\varphi, T\psi) \qquad \psi \in D(T') = D(T)$$

We will often use the foregoing results adapted in this manner to lower semi-bounded forms.

Proposition 1.2.14. Let t be a densely defined, closed, lower semi-bounded form and let T be the associated self-adjoint operator. Further let D be a core of t and for every finite dimensional subspace $M \subset D$ define

$$\lambda(M) = \sup_{\psi \in M,\ \|\psi\| = 1} t(\psi)$$

and for every integer $m \geq 1$ define

$$\lambda_m = \inf_{\dim M = m} \lambda(M)$$

It follows that $\lambda_m \to \infty$ as $m \to \infty$ if and only if the spectrum of T consists of discrete eigenvalues of finite multiplicity and in this case the eigenvalues are given in increasing order, repeated according to multiplicity, by the λ_m.

Note that if t_1 and t_2 are two forms of the kind considered in the proposition and λ^1_m, λ^2_m the corresponding numbers defined by the minimax process then $t_1 \geq t_2$ implies that $\lambda^1_m \geq \lambda^2_m$ and in particular if $\lambda^2_m \to \infty$ as $m \to \infty$ then $\lambda^1_m \to \infty$.

In general if T_1 and T_2 are the self-adjoint operators associated with t_1 and t_2 we will write

$$T_1 \geq T_2$$

whenever $t_1 \geq t_2$.

Proposition 1.2.15. Let t and t' be densely defined, closed, lower semi-bounded forms on $\mathcal{H}$ and let T and T' be the associated self-adjoint operators.

1. Let D be a core of t and $\mathcal{F}$ a finite family of orthonormal vectors $\psi \in D$. The following conditions are equivalent

a) $$\sup_{\mathcal{F}} \sum_{\psi \in \mathcal{F}} \exp\{-t(\psi)\} < +\infty$$

b) $$\mathrm{Tr}_{\mathcal{H}}(e^{-T}) < +\infty$$

and if they are satisfied then

$$\mathrm{Tr}_{\mathcal{H}}(e^{-T}) = \sup_{\mathcal{F}} \sum_{\psi \in \mathcal{F}} \exp\{-t(\psi)\}.$$

2. Consequently if $D(t') \subset D(t)$, $t' \geq t$ and $\mathrm{Tr}_{\mathcal{H}}(e^{-T}) < +\infty$ then

$$\mathrm{Tr}_{\mathcal{H}}(e^{-T'}) \leq \mathrm{Tr}_{\mathcal{H}}(e^{-T}).$$

3. Take $0 < \alpha < 1$ and assume $\alpha t + (1-\alpha)t'$ is densely defined. Let $\alpha T \dotplus (1-\alpha)T'$ denote the operator associated with this latter form and assume e^{-T} and $e^{-T'}$ are of trace class. It follows that

$$\mathrm{Tr}_{\mathcal{H}}\left(e^{-(\alpha T \dot{+} (1-\alpha)T')}\right) \leq \mathrm{Tr}_{\mathcal{H}}(e^{-T})^{\alpha}\,\mathrm{Tr}_{\mathcal{H}}(e^{-T'})^{1-\alpha} .$$

This result is a standard tool in statistical mechanics ; part 1. is proved by using 1.2.14. together with the convexity inequality

$$e^{-t(\psi)} \leq (\psi, e^{-T}\psi) .$$

Part 3. follows from part 2. and the Hölder inequality in the form

$$\sum_{\psi\in\mathcal{F}} e^{-\alpha t(\psi) - (1-\alpha)t'(\psi)} \leq \Big(\sum_{\psi\in\mathcal{F}} e^{-t(\psi)}\Big)^{\alpha}\Big(\sum_{\psi\in\mathcal{F}} e^{-t'(\psi)}\Big)^{1-\alpha} .$$

1.2.16. Let us next illustrate some of the above properties in the context of the differential operators discussed in 1.1.13. - 1.1.21.

Let D_1 be the domain introduced in 1.1.13. the set of $\psi \in L^2(0,L)$ which are absolutely continuous on $[0,L]$ with derivative in , $L^2(0,L)$ and define the form S_σ by

$$D(S_\sigma) = D_1$$

$$S_\sigma(\psi) = \int_0^L dx\,\left|\frac{d\psi}{dx}(x)\right|^2 + \sigma_0|\psi(0)|^2 + \sigma_L|\psi(L)|^2 , \quad \psi\in D_1$$

where $\sigma_0, \sigma_L \in R$. Consider first the case $\sigma = 0$, i.e. $\sigma_0 = 0 = \sigma_L$, then we have

$$S_\sigma(\psi) = (P_0\psi, P_0\psi) \qquad \psi\in D(S_0)$$

where P_0 is the differentiation operator defined in 1.1.13. But P_0 is closed on D_1 by 1.1.14. and hence S_0 is closed by 1.2.7.

Clearly for each $\psi \in D(S_0)$ and $\varphi \in D(P_0^* P_0) \subset D(S_0)$ we have

$$S_0(\psi,\varphi) = (\psi, P_0^* P_0 \varphi) .$$

Thus from parts 3. and 1. of Proposition 1.2.6. we see that $S_0 = P_0^* P_0$ is the operator associated with S_0 .

Next consider the general case $\sigma \neq 0$; we first examine the form Σ given by

$$\Sigma(\psi) = \sigma_0|\psi(0)|^2 + \sigma_L|\psi(L)|^2 \qquad \psi\in D(S_\sigma)$$

$$= \int_0^L dx\,\frac{d}{dx}\left[\xi(x)|\psi(x)|^2\right]$$

where $\xi(x) = (\sigma_0 + \sigma_L)x/L - \sigma_0$. Setting $\sigma_m = \max(|\sigma_0|, |\sigma_L|)$ we find by explicitly differentiating and using $|\xi| \leq \sigma_m$, $|d\xi/dx| \leq 2\sigma_m/L$ that

$$|\Sigma(\psi)| \leq \sigma_m \int_0^L dx \left| \frac{d}{dx} |\psi(x)|^2 \right| + \frac{2\sigma_m}{L} \int_0^L dx\, |\psi(x)|^2$$

$$\leq \sigma_m \, \epsilon \, S_0(\psi) + \left(\frac{\sigma_m}{\epsilon} + \frac{2\sigma_m}{L} \right) \|\psi\|^2$$

for each $\epsilon > 0$ i.e. Σ is S_0-bounded with a relative bound $\epsilon\sigma_m$ which can be chosen arbitrarily small. Thus, for each σ S_σ is lower semi-bounded and closed by 1.2.11. and in particular

$$S_\sigma \geq - \left(\frac{2\sigma_m}{L} + \sigma_m^2 \right) \geq - \left(\sigma_m + \frac{1}{L} \right)^2 .$$

It is easily checked that the self-adjoint operator S_σ associated with S_σ is exactly the operator introduced in 1.1.20. (for example use part 3. of 1.2.6.).

Finally let us define S_∞ in the manner of 1.2.9. as the limit of S_σ as $\sigma \to \infty$, i.e. $\sigma_0, \sigma_L \to \infty$. Thus $D(S_\infty)$ is the subset of D_1 such that

$$\sup_{\sigma_0, \sigma_L} \left[S_0(\psi) + \sigma_0 |\psi(0)|^2 + \sigma_L |\psi(L)|^2 \right] < +\infty$$

i.e. the set of $\psi \in D_1$ such that $\psi(0) = 0 = \psi(L)$ and

$$S_\infty(\psi) = \int_0^L dx \left| \frac{d\psi(x)}{dx} \right|^2 , \qquad \psi \in D(S_\infty) .$$

Hence referring to 1.1.13 we see that $D(S_\infty) = D(P_\infty)$ and

$$S_\infty(\psi) = (P_\infty \psi, P_\infty \psi) , \qquad \psi \in D(S_\infty) .$$

Again S_∞ is closed by 1.1.14. and the associated self-adjoint operator is given by

$$S_\infty = P_\infty^* P_\infty .$$

Note that from 1.2.10. we can conclude that the eigenvalues $\lambda_m(\sigma)$ of S_σ tend to those of S_∞ in the limit $\sigma \to \infty$ confirming abstractly the concrete reasoning of 1.1.20. One also has

$$S_0(\psi) \leq S_\sigma(\psi) \leq S_\infty(\psi) , \qquad \sigma_0, \sigma_L \geq 0$$

and

$$S_\sigma(\psi) \leq S_{\sigma'}(\psi) , \qquad \sigma_0 \leq \sigma_0' , \; \sigma_L \leq \sigma_L'$$

in the sense of 1.2.3. Consequently one deduces from 1.2.14. that the eigenvalues of

1.2.3. Consequently one deduces from 1.2.14. that the eigenvalues of the associated operators satisfy

$$\lambda_m(0) \leq \lambda_m(\sigma) \leq \lambda_m(\infty) \qquad , \sigma_0, \sigma_L \geq 0$$

and

$$\lambda_m(\sigma) \leq \lambda_m(\sigma') \qquad , \sigma_0 \leq \sigma_0', \sigma_L \leq \sigma_L' .$$

Of course in the above discussion it is also possible to consider the case where either σ_0 or σ_L is formally set to be $+\infty$ by restricting the domain of S_σ by the appropriate condition $\psi(0) = 0$ or $\psi(L) = 0$.

1.2.17. Let us next consider the analogous discussion of the Laplacian operator in ν dimensions. We consider the setting of 1.1.16. with Λ an open bounded connected set in R^ν and assume that the surface $\partial\Lambda$ of Λ is smooth. In particular we assume that $\partial\Lambda$ is constituted of one or more continuous surfaces with piecewise continuous tangents.

Let $x \in \partial\Lambda \to \sigma(x)$ denote a real piecewise continuous function ; then we define the form S_σ over $L^2(\Lambda)$ by

$$D(S_\sigma) = C^1(\Lambda)$$

$$S_\sigma(\psi) = \int_\Lambda dx\, |\underline{\nabla}_x \psi(x)|^2 + \int_{\partial\Lambda} dS\, \sigma(x)\, |\psi(x)|^2 \quad , \psi \in C^1(\Lambda).$$

Again considering the case $\sigma = 0$ first we can conclude that S_σ is a positive closable form because

$$S_\sigma(\psi) = (\underline{P}_0 \psi , \underline{P}_0 \psi) \qquad , \quad \psi \in C^1(\Lambda) = D(\underline{P}_0)$$

where $\underline{P}_0$ is the vector-valued operator of 1.1.16. But in this latter section we have shown that $\underline{P}_0$ is closable and hence the closability of S_σ follows from 1.2.7.

The case $\sigma \neq 0$ is analogous to the 1-dimensional problem if we assume that σ has a continuous extension $x \in \Lambda \longrightarrow \sigma(x) \in C^1(\Lambda)$. In this case we can introduce a vector-valued function $\underline{\xi}(x) \in (C^1(\Lambda))^\nu$ such that

$$\int_{\partial\Lambda} dS\, \sigma(x)\, |\psi(x)|^2 = \int_\Lambda dx\, \underline{\nabla}_x . \left(\underline{\xi}(x)\, \sigma(x)\, |\psi(x)|^2 \right)$$

and then follow the estimation procedure of the previous paragraph to conclude that

$$\left| \int_{\partial\Lambda} dS\, \sigma(x) |\psi(x)|^2 \right| \leq \epsilon\, |\sigma|\, |\underline{\xi}|\, S_0(\psi) + \left(\frac{|\underline{\xi}||\sigma|}{\epsilon} + |\underline{\nabla}.(\underline{\xi}\sigma)| \right) \|\psi\|^2$$

for each $\epsilon > 0$ where

$$|\sigma| = \sup_{x \in \Lambda} |\sigma(x)| \quad , \quad |\xi| = \sup_{x \in \Lambda} \sum_{i=1}^{\nu} |\xi_i(x)| \qquad \text{etc.}$$

We can conclude as before that S_σ is closable.

Further if we take a family of functions σ_α such that $\sigma_\alpha \geq \alpha \in R$ and define

$$S_\infty(\psi) = \lim_{\alpha \to \infty} S_{\sigma_\alpha}(\psi) = \sup_\alpha S_{\sigma_\alpha}(\psi)$$

then the domain of S_∞ is exactly $C_0^1(\Lambda)$ and on this domain

$$S_\infty(\psi) = (\underline{P}_\infty \psi, . \underline{P}_\infty \psi) \quad ;$$

thus S_∞ is closable due to the closability of $\underline{P}_\infty$.

The above forms satisfy similar inequalities to those given in the 1-dimensional case with similar consequences for the eigenvalues of the associated self-adjoint operators.

If Λ is a parallelepiped we can also define a form $S(\theta)$ by

$$(S(\theta))(\psi) = (\underline{P}(\theta)\psi, . \underline{P}(\theta)\psi) \quad , \qquad \psi \in D(\underline{P}(\theta))$$

and one has

$$S_0 \leq S(\theta) \leq S_\infty$$

The self-adjoint operators associated with the above forms correspond to different self-adjoint extensions on $L^2(\Lambda)$ of the formal differential operator $-\underline{\nabla}_x^2$ with the following boundary conditions

$$\frac{\partial \psi}{\partial n} - \sigma\psi = 0 \qquad \text{on} \quad \partial\Lambda \quad \text{for} \quad S_\sigma$$

$$\psi = 0 \qquad \text{on} \quad \partial\Lambda \quad \text{for} \quad S_\infty$$

and

$$\left.\begin{aligned} \psi(x^{(1)}, \ldots, 0, \ldots, x^{(\nu)}) &= e^{i\theta_i} \psi(x^{(1)}, \ldots, L^{(i)}, \ldots, x^{(\nu)}) \\ \frac{\partial \psi}{\partial x^{(i)}}(x^{(1)}, \ldots, 0, \ldots, x^{(\nu)}) &= \cdot e^{i\theta_i} \frac{\partial \psi}{\partial x^{(i)}}(x^{(1)}, \ldots, L^{(i)}, \ldots, x^{(\nu)}) \end{aligned}\right\} \text{for} \quad S(\theta)$$

where $\partial/\partial n$ denotes the inward normal derivative. However a function in the domain of these operators is not necessarily differentiable in the normal sense and all the differential relations, including the boundary conditions, must be interpreted in the sense of distributions. It was to avoid these technical difficulties that we chose to describe the operators implicitly by closable forms and did not give a detailed description of the domains of their closures.

§3 - LOCAL HAMILTONIANS

1.3.1. The methods of §2 can now be used to characterize the operators corresponding to Hamiltonians of finite systems of particles. We will explicitly consider particles satisfying Bose statistics both in the canonical and grand canonical schemes, i.e. with a fixed number of particles and with a varying number. The description of Fermi particles is very similar. We first describe the kinetic energy operators in terms of lower semi-bounded forms and then adopt the position that the interaction operator is describable in the same manner. The total Hamiltonian is taken to be the generalized sum (cf. 1.2.12.) of the two operators.

1.3.2. The Hilbert space $\mathcal{H}^{(n)}(\Lambda)$ appropriate to the description of n Bose particles confined to the open bounded region Λ of R^ν is the space of completely symmetric, complex, square integrable, functions of n points in Λ with the scalar product

$$(\varphi, \psi)_n = \int_{\Lambda^n} \frac{dx_1 \ldots dx_n}{n!} \overline{\varphi(x_1, \ldots, x_n)}\, \psi(x_1, \ldots, x_n) .$$

We adopt a set theoretic notation writing $X_n = \{x_1, \ldots, x_n\}$, $\psi(X_n) = \psi(x_1, \ldots, x_n)$ and defining the measure dX_n by

$$\int_\Lambda dX_n \cdot = \int_{\Lambda^n} \frac{dx_1 \ldots dx_n}{n!} .$$

Thus the scalar product is given by

$$(\varphi, \psi)_n = \int_\Lambda dX_n\, \overline{\varphi(X_n)}\, \psi(X_n) .$$

The Fock space $\mathcal{H}(\Lambda)$, the Hilbert space used to describe an arbitrary number of particles in Λ, is defined by the direct sum

$$\mathcal{H}(\Lambda) = \bigoplus_{n \geq 0} \mathcal{H}^{(n)}(\Lambda)$$

where we take $\mathcal{H}^0(\Lambda) = \mathbb{C}$. Thus an element of $\mathcal{H}(\Lambda)$ can be considered as a function over the finite sets $X \subset \Lambda$ with scalar product

$$(\varphi, \psi) = \int_\Lambda dX\, \overline{\varphi(X)}\, \psi(X)$$

where

$$\int_\Lambda dX \cdot = \sum_{n \geq 0} \frac{1}{n!} \int_{\Lambda^n} dx_1 \ldots dx_n .$$

and

$$\psi(X) = \psi(x_1, \ldots, x_n) \qquad \text{if } X = \{x_1, \ldots, x_n\}.$$

1.3.3. For later use we will need connections between the above Hilbert spaces for different Λ . If Λ_1 and Λ_2 are disjoint we have

$$\|\psi\|_n^2 = \sum_{m=0}^{n} \int_{\Lambda_1} dX_m \int_{\Lambda_2} dY_{n-m} \, |\psi(X_m \cup Y_{n-m})|^2 \quad , \psi \in \mathcal{H}^{(n)}(\Lambda_1 \cup \Lambda_2).$$

Now $(X_m , Y_{n-m}) \to \psi(X_m \cup Y_{n-m})$ can be identified as an element of the tensor product space $\mathcal{H}^{(m)}(\Lambda_1) \otimes \mathcal{H}^{(n-m)}(\Lambda_2)$ and with this identification we have

$$\mathcal{H}^{(n)}(\Lambda_1 \cup \Lambda_2) = \bigoplus_{m=0}^{n} \mathcal{H}^{(m)}(\Lambda_1) \otimes \mathcal{H}^{(n-m)}(\Lambda_2)$$

(cf. 1.1.2. and 1.1.3.). Further we have

$$\begin{aligned} \mathcal{H}(\Lambda_1 \cup \Lambda_2) &= \bigoplus_{n \geq 0} \bigoplus_{m=0}^{n} \mathcal{H}^{(m)}(\Lambda_1) \otimes \mathcal{H}^{(n-m)}(\Lambda_2) \\ &= \mathcal{H}(\Lambda_1) \otimes \mathcal{H}(\Lambda_2) . \end{aligned}$$

as can be verified directly.

1.3.4. From now on we assume that each Λ under consideration is connected and has a smooth boundary. We define the following forms over $\mathcal{H}^{(n)}(\Lambda)$ for each $n \geq 0$,

$$t^{(n)}_{\infty,\Lambda} : \quad \begin{aligned} D(t^{(n)}_{\infty,\Lambda}) &= C_0^1(\Lambda^n) \quad , \ n \geq 1 \\ t^{(n)}_{\infty,\Lambda}(\psi) &= \int_{\Lambda} dX_n \sum_{x \in X_n} |\underline{\nabla}_x \psi(X_n)|^2 \end{aligned}$$

$$t^{(n)}_{\sigma,\Lambda} : \quad \begin{aligned} D(t^{(n)}_{\sigma,\Lambda}) &= C^1(\Lambda^n) \quad , \ n \geq 1 \\ t^{(n)}_{\sigma,\Lambda}(\psi) &= \int_{\Lambda} dX_n \sum_{x \in X_n} |\underline{\nabla}_x \psi(X_n)|^2 + \int_{\partial\Lambda} dS_n \, \sigma(X_n) \, |\psi(X_n)|^2 \end{aligned}$$

$$t^{(0)}_{\infty,\Lambda} = 0$$

$$t^{(0)}_{\sigma,\Lambda} = 0$$

where σ is a real, completely symmetric, continuous function over the surface of Λ^n and

$$\int_{\partial\Lambda} dS_n \; \sigma(x_n) |\psi(x_n)|^2 = \frac{1}{n!} \sum_{i=1}^{n} \int_{\Lambda} dx_1 \ldots \int_{\partial\Lambda} dS_i \ldots \int_{\Lambda} dx_n \; \sigma(x_1 \ldots x_n) |\psi(x_1 \ldots x_n)|^2 .$$

In fact we will restrict our attention to the simple case that σ is a constant : it should be noted that for the properties we consider in the next chapter this leads to no loss of generality. Thus σ will be assumed constant from henceforth.

The forms defined above are closable and lower semi-bounded by the same arguments as used in 1.2.17. We denote by $T^{(n)}_{\infty,\Lambda}$ and $T^{(n)}_{\sigma,\Lambda}$ the self-adjoint operators associated with their respective closures. Following the reasoning of 1.1.20. and 1.2.17. one sees that these operators correspond to the kinetic energy operators of Bose particles in a "box" Λ with repulsive walls and walls of "elasticity" σ respectively.

The corresponding kinetic energy operators on $\mathcal{H}(\Lambda)$ are defined as the direct sums (cf. 1.1.12.).

$$T_{\infty,\Lambda} = \bigoplus_{n\geq 0} T^{(n)}_{\infty,\Lambda}$$

$$T_{\sigma,\Lambda} = \bigoplus_{n\geq 0} T^{(n)}_{\sigma,\Lambda} .$$

Note that $T_{\sigma,\Lambda}$ is not necessarily lower semi-bounded ; this is the case if $\sigma \geq 0$ but if $\sigma < 0$ one can only conclude that $t^{(n)}_{\sigma,\Lambda}$ has a bound of the form

$$t^{(n)}_{\sigma,\Lambda} \geq n \, \lambda_1(\sigma) \, \|\psi\|^2$$

where $\lambda_1(\sigma)$ is the lowest eigenvalue of $T^{(1)}_{\sigma,\Lambda}$. The factor n then destroys the lower-semiboundedness of $T_{\sigma,\Lambda}$. Thus one cannot generally characterize these latter operators by lower-semibounded forms. In this latter case it is useful to define the operators

$$K^{\mu}_{\sigma,\Lambda} = \bigoplus_{n\geq 0} \left[T^{(n)}_{\sigma,\Lambda} - \mu n \, 1^{(n)}_{\Lambda} \right]$$

where $1^{(n)}_{\Lambda}$ is the identity operator on $\mathcal{H}^{(n)}(\Lambda)$; if $\mu < \lambda_1(\sigma)$ then $K^{\mu}_{\sigma,\Lambda}$ is positive.

1.3.5. To describe interactions which conserve the number of particles we can postulate that the interaction is mediated by a lower semi-bounded self-adjoint operator $U_\Lambda^{(n)}$ acting on $\mathcal{H}^{(n)}(\Lambda)$ with domain $D(U_\Lambda^{(n)})$. Corresponding to each such operator there is a closed lower semi-bounded form $u_\Lambda^{(n)}$ defined by closing the form $\psi \in D(U_\Lambda^{(n)}) \to (\psi, U_\Lambda^{(n)}\psi)$. The total Hamiltonians describing the n interacting particles in Λ can now be defined as the generalized sum (cf. 1.2.12.) of $U_\Lambda^{(n)}$ and the appropriate kinetic energy operators.

To ensure the possibility of this definition we assume that the domains $D(t_{\infty,\Lambda}^{(n)}) \cap D(u_\Lambda^{(n)})$ and $D(t_{\sigma,\Lambda}^{(n)}) \cap D(u_\Lambda^{(n)})$ are dense in $\mathcal{H}^{(n)}(\Lambda)$ and define

$$h_{\infty,\Lambda}^{(n)} : \quad \begin{aligned} D(h_{\infty,\Lambda}^{(n)}) &= D(t_{\infty,\Lambda}^{(n)}) \cap D(u_\Lambda^{(n)}) \\ h_{\infty,\Lambda}^{(n)} &= t_{\infty,\Lambda}^{(n)} + u_\Lambda^{(n)} \end{aligned}$$

$$h_{\sigma,\Lambda}^{(n)} : \quad \begin{aligned} D(h_{\sigma,\Lambda}^{(n)}) &= D(t_{\sigma,\Lambda}^{(n)}) \cap D(u_\Lambda^{(n)}) \\ h_{\sigma,\Lambda}^{(n)} &= t_{\sigma,\Lambda}^{(n)} + u_\Lambda^{(n)} \end{aligned} \quad .$$

These forms are then closable and lower semi-bounded. We denote by $H_{\infty,\Lambda}^{(n)}$ and $H_{\sigma,\Lambda}^{(n)}$ the self-adjoint operators associated with their closures. Thus

$$H_{\infty,\Lambda}^{(n)} = T_{\infty,\Lambda}^{(n)} \dot{+} U_\Lambda^{(n)} \qquad \text{etc.}$$

The corresponding Hamiltonians on $\mathcal{H}(\Lambda)$ are again defined as direct sum

$$H_{\infty,\Lambda} = \bigoplus_{n\geq 0} H_{\infty,\Lambda}^{(n)} \quad , \quad H_{\sigma,\Lambda} = \bigoplus_{n\geq 0} H_{\sigma,\Lambda}^{(n)} \quad .$$

1.3.6. In the special case that Λ is a parallelepiped we can use the material of 1.2.17. and the same general scheme as above to define Hamiltonians corresponding to periodic boundary conditions. We will not introduce these Hamiltonians explicitly as they will play a minor role in the discussion of chapter II .

GENERAL BIBLIOGRAPHY

For properties of differential operators one may consult

E. CODDINGTON and N. LEVINSON

Theory of Ordinary Differential Operators - Mc Graw-Hill (New York) 1955

M.A. NEUMARK

Lineare Differentialoperatoren - Akademic (Berlin) 1960

The general theory of operators on Hilbert space and sesquilinear forms is given in :

T. KATO

Perturbation Theory for Linear Operators - S Springer (Berlin) 1966

F. RIESZ and B. NAGY

Leçons d'Analyse Fonctionnelle - Gauthier-Villars (Paris) 1965

All self-adjoint extensions of first and second order differential operators are characterized at length in

M.H. STONE

Linear Transformations in Hilbert Space - American Math.Soc. (New York)

EXERCISES

1. a. Prove that the direct sum space defined in 1.1.2. is complete.

   b. Prove that the infinite sum of positive closed (closable) forms defined in 1.2.9. is closed (closable).

2. Repeat the discussion of the spectra of S_σ , given in 1.1.20, for the case $\sigma_0 \leq 0$, $\sigma_L > 0$ and deduce in particular that there is only one negative eigenvalue (in physical terms if both walls are attractive, $\sigma_0 < 0$, $\sigma_L < 0$ there are two possible bound states but if only one wall is attractive, $\sigma_0 < 0$, $\sigma_L > 0$ there is only one bound state).

3. Take $\sigma_0 = \sigma_L = \sigma$ and use the estimate of 1.2.16. to show that for $\sigma_1 \geqslant \sigma_2$ in an interval $\Delta \subset R$ there exist a_Δ , $b_\Delta > 0$ such that the eigenvalues $\lambda_n(\sigma)$ of S_σ satisfy

$$0 \leq \lambda_n(\sigma_1) - \lambda_n(\sigma_2) \leq (a_\Delta + \lambda_n(\sigma_2))(e^{b_\Delta(\sigma_1-\sigma_2)} - 1)$$

[<u>Hint</u> : Using the estimate of Σ_1 given in 1.2.16. one finds that there are positive constants c_Δ , d_Δ such that

$$S_{\sigma_2} \leq S_{\sigma_1} \leq S_{\sigma_2} + c_\Delta (\sigma_1 - \sigma_2) S_{\sigma_2} + d_\Delta (\sigma_1 - \sigma_2) .$$

Hence from 1.2.14. one concludes that

$$\lambda_n(\sigma_2) \leq \lambda_n(\sigma_1) \leq \lambda_n(\sigma_2) + c_\Delta (\sigma_1 - \sigma_2) \lambda_n(\sigma_2) + d_\Delta (\sigma_1 - \sigma_2) .$$

Whence the differential inequality

$$0 \leq \frac{d\lambda_n(\sigma)}{d\sigma} \leq c_\Delta \lambda_n(\sigma) + d_\Delta \qquad , \quad \sigma \in \Delta .$$

Solving this inequality gives the desired answer.]

4. Prove that with $\sigma_0 = \sigma_L = \sigma$

$$\begin{aligned} Tr(e^{-S_\sigma}) &\leq 2e^{\sigma^2 + 2|\sigma|/L} + 1 + \frac{L}{2\sqrt{\pi}} \\ &= q(L, \sigma) \end{aligned}$$

where the first term on the right hand side can be omitted if $\sigma \geq 0$ and

$$Tr(e^{-S_\infty}) \leq \frac{L}{2\sqrt{\pi}} .$$

In the multi-dimensional case with Λ a parallepiped with edges of length $L^{(1)}, .., L^{(\nu)}$ and S_σ the self-adjoint operator associated with the form S_σ deduce that

$$Tr(e^{-S_\sigma}) \leq \prod_{i=1}^{\nu} q(L^{(i)}, \sigma) .$$

CHAPTER II

THE THERMODYNAMIC PRESSURE

INTRODUCTION

In this second chapter we will prove the existence of the thermodynamic pressure as a limit of the pressures of finite systems in the grand canonical ensemble. We will restrict our discussion to positive decreasing interactions and show that the thermodynamic pressure is independent of the boundary conditions used to describe the finite systems which enter into its definition.

Our discussion of the interacting case is based upon the properties of the pressure of the non-interacting system which we study in section §1 . In section §2 we show that the results obtained for the non-interacting case are easily extended to positive decreasing interactions. We attempt to give a relatively complete discussion of the various boundedness and sub-additivity properties of the partition functions of the non-interacting system with the hope that this will be of use in extending the present results to a more general class of interactions. In section §3 we introduce finite range interactions and give a number of results concerning the free energy and periodized systems.

§1 - THE NON-INTERACTING SYSTEMS

2.1.1. The pressure of a finite system of Bosons, confined to the region $\Lambda \subset R^\nu$ with elastic or repulsive boundary conditions, is defined in terms of the canonical partition functions

$$Q^{\sigma}_{\Lambda}(\beta, n) = \mathrm{Tr}_{\mathcal{H}^{(n)}(\Lambda)}\left(e^{-\beta T^{(n)}_{\sigma,\Lambda}}\right) \quad , \quad \beta > 0$$

$$Q^{\infty}_{\Lambda}(\beta, n) = \mathrm{Tr}_{\mathcal{H}^{(n)}(\Lambda)}\left(e^{-\beta T^{(n)}_{\infty,\Lambda}}\right) \quad , \quad \beta > 0$$

or the grand canonical partition functions

$$Z^{\sigma}_{\Lambda}(\beta, \mu) = \sum_{n \geq 0} e^{\beta\mu n} Q^{\sigma}_{\Lambda}(\beta, n)$$

$$Z^{\infty}_{\Lambda}(\beta, \mu) = \sum_{n \geq 0} e^{\beta\mu n} Q^{\infty}_{\Lambda}(\beta, n)$$

where $T^{(n)}_{\sigma,\Lambda}$ and $T^{(n)}_{\infty,\Lambda}$ are the Hamiltonians introduced in 1.3.4. The parameter μ is interpretable as the chemical potential : we will often replace it by the variable $z = e^{\beta\mu}$.

Suitably interpreting the trace, each of these functions is given as an infinite sum of positive terms and thus the definition is unambigous. However one is principally interested in these functions when they are finite-valued and our first aim is to calculate bounds for them. Note that

$$0 \leq Q^{\sigma}_{\Lambda}(\beta, n) \leq z^{-n} Z^{\sigma}_{\Lambda}(\beta, z) \qquad \text{etc.}$$

and hence it suffices to calculate an upper bound for the grand canonical partition functions.

This is very simple if Λ is a parallelepiped and this will be assumed from now on, unless the contrary is explicitly stated.

We first introduce a notation which will be constantly used in the sequel. If Λ is a parallelepiped with edges of length $L^{(1)}, \ldots, L^{(\nu)}$ we denote its volume by $V(\Lambda)$, i.e.

$$V(\Lambda) = \prod_{i=1}^{\nu} L^{(i)}$$

and introduce a measure L of its linear dimension by

$$L = \min_{1 \leq i \leq \nu} L^{(i)} .$$

<u>Proposition 2.1.2.</u> Let Λ be a parallelepiped. If $\mu < \mu(\sigma)$ where

$$\mu(\sigma) = 0 \quad , \quad \sigma > 0$$

$$= -\nu|\sigma|\left(|\sigma| + \frac{2}{L}\right) \quad , \quad \sigma \leq 0$$

then

$$Z^{\sigma}_{\Lambda}(\beta, z) \leq \exp\left\{\frac{z}{1 - z e^{\beta\mu(\sigma)}} \frac{V(\Lambda)}{(4\pi\beta)^{\nu/2}} \left[1 + \epsilon(\sigma, L)\right]\right\}$$

where $\epsilon(\sigma, L) \to 0$ as $L \to \infty$. The same bound is valid for $Z^{\infty}_{\Lambda}(\beta, z)$ for all $\beta > 0$, $\mu < 0$ and with $\mu(\sigma) = 0 = \epsilon(\sigma, L)$.

Let $\{\lambda_n(\sigma)\}_{n \geq 1}$ denote the eigenvalues of $T^{(1)}_{\sigma,\Lambda}$ arranged in increasing order and repeated according to multiplicity. If $\mu < \lambda_1(\sigma)$ a standard calculation gives

$$\begin{aligned}
Z^{\sigma}_{\Lambda}(\beta, z) &= \prod_{n \geq 1} \frac{1}{1 - z e^{-\beta\lambda_n(\sigma)}} , \quad \beta > 0 \\
&= \prod_{n \geq 1} \left\{1 + \frac{z e^{-\beta\lambda_n(\sigma)}}{1 - z e^{-\beta\lambda_n(\sigma)}}\right\} \\
&\leq \prod_{n \geq 1} \exp\left\{\frac{z e^{-\beta\lambda_n(\sigma)}}{1 - z e^{-\beta\lambda_n(\sigma)}}\right\} \\
&\leq \exp\left\{\frac{z}{1 - z e^{-\beta\lambda_1(\sigma)}} \sum_{n \geq 1} e^{-\beta\lambda_n(\sigma)}\right\}.
\end{aligned}$$

But the estimate for S_σ given in 1.2.16. and 1.2.14. show that

$$\lambda_1(\sigma) \geq -\sum_{i=1}^{\nu} |\sigma|\left(|\sigma| + \frac{2}{L^{(i)}}\right) \geq -\nu|\sigma|\left(|\sigma| + \frac{2}{L}\right)$$

if $\sigma < 0$ and $\lambda_1(\sigma) \geq 0$ if $\sigma \geq 0$. Finally from exercise 4. of chapter I we have

$$\sum_{n \geq 1} e^{-\beta\lambda_n(\sigma)} \leq \frac{V(\Lambda)}{(4\pi\beta)^{\nu/2}}\left(1 + \epsilon(\sigma, L)\right)$$

with

$$\begin{aligned}
\epsilon(\sigma, L) &= \left[1 + \frac{\sqrt{4\pi\beta}}{L}\left(1 + 2e^{-\beta\lambda_1(\sigma)/\nu}\right)\right]^{\nu} - 1 , \quad \sigma < 0 \\
&= \left[1 + \frac{\sqrt{4\pi\beta}}{L}\right]^{\nu} - 1 , \quad \sigma \geq 0.
\end{aligned}$$

The estimate for $Z^{\infty}_{\Lambda}(\beta, z)$ is similar.

As we are eventually interested in large systems, e.g. large parallelepipeds, we will now assume throughout that the parallelepipeds we consider are such that L is greater than some fixed value L_0. We also introduce $D_\sigma \subset R^2$

to be the range of the thermodynamic parameters (β,μ) i.e. $D_\sigma = \{(\beta,\mu)\,;\ \beta>0,\ \mu<\mu(\sigma)\}$.

Proposition 2.1.3. The function (pressure)

$$(\beta,\mu)\in D_\sigma \longrightarrow P^\sigma_\Lambda(\beta,\mu) = \frac{1}{V(\Lambda)}\operatorname{Log} Z^\sigma_\Lambda(\beta,\mu)$$

is convex in β and μ and is an increasing function of μ.

The family of functions $(\beta,\mu)\in D_\sigma \to P^\sigma_\Lambda(\beta,\mu)$ is equicontinuous, i.e. given $(\beta,\mu)\in D_\sigma$ and $\epsilon>0$ there is a neighbourhood $N_{\beta,\mu}\subset D_\sigma$ such that $(\beta_1,\mu_1)\in N_{\beta,\mu}$ implies

$$\left| P^\sigma_\Lambda(\beta_1,\mu_1) - P^\sigma_\Lambda(\beta,\mu)\right| < \epsilon$$

uniformly in $L^{(1)},\ldots,L^{(\nu)} > L_0$.

Similar results are valid for $P^\infty_\Lambda(\beta,\mu) = V(\Lambda)^{-1}\operatorname{Log} Z^\infty_\Lambda(\beta,\mu)$.

First note that from part 3 of 1.2.15 one deduces that for $\alpha\in[0,1]$

$$Q^\sigma_\Lambda\left(\alpha\beta_1+(1-\alpha)\beta_2,\, n\right) \le Q^\sigma_\Lambda(\beta_1,n)^\alpha\, Q^\sigma_\Lambda(\beta_2,n)^{1-\alpha}$$

and consequently

$$Z^\sigma_\Lambda\left(\alpha\beta_1+(1-\alpha)\beta_2,\, z\right) \le \sum_{n\ge 0}\left(z^n Q^\sigma_\Lambda(\beta_1,n)\right)^\alpha \left(z^n Q^\sigma_\Lambda(\beta_2,n)\right)^{1-\alpha}$$

$$\le Z^\sigma_\Lambda(\beta_1,z)^\alpha\, Z^\sigma_\Lambda(\beta_2,z)^{1-\alpha}$$

where the last step uses the Hölder inequality and $Q_\Lambda \ge 0$. The convexity of P_Λ in β follows immediately.

The convexity in μ follows from a similar application of the Hölder inequality and the increase property in μ follows because the $Q_\Lambda \ge 0$ and the logarithm is monotonically increasing.

The equicontinuity property follows from the convexity properties and the bounds derived in 2.1.2. Explicitly one calculates that if $x>0 \to f(x)\ge 0$ is a non-negative convex function and $h\ge 0$, $1>a>0$ and $b>0$ then

$$-\frac{h}{ax} f((1-a)x) \le f(x+h)-f(x) \le \frac{h}{bx} f((1+b)x).$$

Applying this formula twice to $(\beta,\mu)\longrightarrow P^\sigma_\Lambda(\beta,\mu)$ and then using the fact that P^σ_Λ is bounded uniformly in $L^{(1)},\ldots,L^{(\nu)} > L_0$ (2.1.2.) one obtains the desired result.

It is also possible to derive similar properties of the pressure as a function of σ but as this parameter does not have the same fundamental thermodynamic significance we have chosen to separate the statement.

Proposition 2.1.4. For fixed $(\beta,\mu)\in D_\sigma$ the function $\sigma\to P^\sigma_\Lambda(\beta,\mu)$ is convex, and consequently continuous. Further

$$P^\infty_\Lambda(\beta,\mu) = \lim_{\sigma\to\infty} P^\sigma_\Lambda(\beta,\mu).$$

The proof of convexity follows again from 1.2.15. when one notes that

$$T^{(n)}_{\alpha\sigma_1+(1-\alpha)\sigma_2,\Lambda} = \alpha T^{(n)}_{\sigma_1,\Lambda} \dot{+} (1-\alpha)T^{(n)}_{\sigma_2,\Lambda}.$$

The last statement follows because the eigenvalues $\{\lambda_n(\sigma)\}_{n\geq 1}$ of $T^{(1)}_{\sigma,\Lambda}$ converge to the eigenvalues of $T^{(1)}_{\infty,\Lambda}$ as $\sigma\to\infty$ (cf. the discussion of 1.1.20 for the 1-dimensional case) as a result of 1.2.10.

Next we examine the behaviour of Q_Λ and Z_Λ as Λ varies. There are two cases which are particularly easy to describe and which can be roughly described as opposite extremes. The first, and easiest to manage, is the case $\sigma=+\infty$ and the second $\sigma=0$.

Proposition 2.1.5. (Ruelle-Fisher) Let Λ_1 and Λ_2 be arbitrary disjoint open bounded sets in R^ν then

$$Q^\infty_{\Lambda_1\cup\Lambda_2}(\beta,n) \geq \sum_{m=0}^{n} Q^\infty_{\Lambda_1}(\beta,m)\,Q^\infty_{\Lambda_2}(\beta,n-m),\quad \beta>0.$$

Consequently

$$Z^\infty_{\Lambda_1\cup\Lambda_2}(\beta,\mu) \geq Z^\infty_{\Lambda_1}(\beta,\mu)\,Z^\infty_{\Lambda_2}(\beta,\mu),\quad \beta>0,\ \mu<0.$$

The functions $\Lambda\to Q^\infty_\Lambda(\beta,n)$, $\Lambda\to Z^\infty_\Lambda(\beta,\mu)$ are increasing, i.e. if $\Lambda\subset\Lambda'$ then $Q^\infty_\Lambda(\beta,n)\leq Q^\infty_{\Lambda'}(\beta,n)$.

The proof is based upon 1.2.15. and the structure described in 1.3.3. Let $\mathcal{F}^m_1$ and $\mathcal{F}^{n-m}_2$, $m=0,1,\ldots,n$ denote finite families of orthonormal functions $\psi^m_1\in D(t^{(m)}_{\infty,\Lambda_1})$, $\psi^{n-m}_2\in D(t^{(n-m)}_{\infty,\Lambda_2})$ respectively. Then $\{\psi^m_1\,\psi^{n-m}_2\}_{n\geq m\geq 0}$ can be identified with a function $\psi^n_{12}\in D(t^{(n)}_{\infty,\Lambda_1\cup\Lambda_2})$ by setting

$$\psi^n_{12}(X_m\cup Y_{n-m}) = \psi^m_1(X_m)\,\psi^{n-m}_2(Y_{n-m}),\quad X_m\subset\Lambda_1,\ Y_{n-m}\subset\Lambda_2.$$

The family $\mathcal{G}^n_{12}$ of functions formed in this manner from the $\mathcal{F}^m_1$, $\mathcal{F}^{n-m}_2$

is a special finite family of orthonormal functions in $D(t^{(n)}_{\infty,\Lambda_1\cup\Lambda_2})$. For $\psi^n_{12} = \{\psi^m_1 \psi^{n-m}_2\}_{n\geq m\geq 0} \in \mathcal{G}^n_{12}$ one has by direct calculation

$$t^{(n)}_{\infty,\Lambda_1\cup\Lambda_2}(\psi^n_{12}) = \sum_{m=0}^{n} t^{(m)}_{\infty,\Lambda_1}(\psi^m_1) + t^{(n-m)}_{\infty,\Lambda_2}(\psi^{n-m}_2).$$

Thus using 1.2.15 one finds

$$\mathrm{Tr}_{\mathcal{H}^{(n)}(\Lambda_1\cup\Lambda_2)}\left(e^{-\beta T^{(n)}_{\infty,\Lambda_1\cup\Lambda_2}}\right) = Q^{\infty}_{\Lambda_1\cup\Lambda_2}(\beta,n)$$

$$\geq \sup_{\mathcal{G}^n_{12}} \sum_{\psi^n_{12}\in\mathcal{G}^n_{12}} e^{-\beta t^{(n)}_{\infty,\Lambda_1\cup\Lambda_2}(\psi^n_{12})}$$

$$= \sup_{\mathcal{G}^n_{12}} \sum_{m=0}^{n}\left(\sum_{\psi^m_1\in\mathcal{F}^m_1} e^{-\beta t^{(m)}_{\infty,\Lambda_1}(\psi^m_1)}\right)\left(\sum_{\psi^{n-m}_2\in\mathcal{F}^{n-m}_2} e^{-\beta t^{(n-m)}_{\infty,\Lambda_2}(\psi^{n-m}_2)}\right)$$

$$= \sum_{m=0}^{n} Q^{\infty}_{\Lambda_1}(\beta,m)\, Q^{\infty}_{\Lambda_2}(\beta,n-m).$$

Clearly

$$Z^{\infty}_{\Lambda_1\cup\Lambda_2}(\beta,z) \geq \sum_{n=0}^{\infty} z^n \sum_{m=0}^{n} Q^{\infty}_{\Lambda_1}(\beta,m)\, Q^{\infty}_{\Lambda_2}(\beta,n-m)$$

$$= Z^{\infty}_{\Lambda_1}(\beta,z)\, Z^{\infty}_{\Lambda_2}(\beta,z).$$

The increase property of Q_Λ and Z_Λ also follows from 1.2.15. because the finite families $\mathcal{F}$ of orthonormal functions $\psi \in D(t^{(n)}_{\infty,\Lambda'})$ which are such that $\psi(X_n) = 0$ whenever $X_n \not\subset \Lambda$ are a special family and for each such function

$$t^{(n)}_{\infty,\Lambda'}(\psi) = t^{(n)}_{\infty,\Lambda}(\psi).$$

Proposition 2.1.6. Let Λ_1 and Λ_2 be arbitrary disjoint open bounded sets in R^ν with smooth boundaries and such that $\Lambda_1\cup\Lambda_2$ has a smooth boundary then

$$Q^{o}_{\Lambda_1\cup\Lambda_2}(\beta,n) \leq \sum_{m=0}^{n} Q^{o}_{\Lambda_1}(\beta,m)\, Q^{o}_{\Lambda_2}(\beta,n-m), \quad \beta>0.$$

Consequently

$$Z^{o}_{\Lambda_1\cup\Lambda_2}(\beta,\mu) \leq Z^{o}_{\Lambda_1}(\beta,\mu)\, Z^{o}_{\Lambda_2}(\beta,\mu), \quad \beta>0,\ \mu<0.$$

If Λ is a parallelepiped then the functions $\Lambda \to Q^{o}_{\Lambda}(\beta,n)$, $\Lambda \to Z^{o}_{\Lambda}(\beta,\mu)$ are increasing functions of the edges $L^{(1)},\dots,L^{(\nu)}$ of Λ.

The proof of the first part of the proposition is again based upon the use of 1.2.15. but uses more of the product structure of $\mathcal{H}^{(n)}(\Lambda_1 \cup \Lambda_2)$ First note that if $\psi \in D(t_{0,\Lambda}^{(n)})$ we can write

$$t_{0,\Lambda_1\cup\Lambda_2}^{(n)}(\psi) = \sum_{m=0}^{n} s_{0,\Lambda_1}^{(m)}(\psi) + s_{0,\Lambda_2}^{(n-m)}(\psi)$$

where $s_{0,\Lambda_1}^{(m)}$ and $s_{0,\Lambda_2}^{(n-m)}$ are defined by

$$s_{0,\Lambda_1}^{(m)}(\psi) = \int_{\Lambda_1} dX_m \int_{\Lambda_2} dY_{n-m} \sum_{x\in X_m} |\nabla_x \psi(X_m \cup Y_{n-m})|^2$$

$$s_{0,\Lambda_2}^{(n-m)}(\psi) = \int_{\Lambda_1} dX_m \int_{\Lambda_2} dY_{n-m} \sum_{y\in Y_{n-m}} |\nabla_y \psi(X_m \cup Y_{n-m})|^2 .$$

Next let us define extensions $\hat{s}_{\Lambda_1}^{(m)}$ etc. of each of these forms by adding to their domains the sets of functions $C^1(\Lambda_1^m) \otimes \mathcal{H}^{(n-m)}(\Lambda_2)$etc. The resulting forms are positive, densely defined, and closable. The last property follows by the same argument as used in 1.2.17. for s_0 , and is based upon the closability of a suitably defined vector-valued differentiation operator the analogue of $\underline{P}_0$ of 1.1.16. But $\hat{s}_{\Lambda_1}^{(m)}$ is also an extension of the form $r_{\Lambda_1}^{(m)}$ defined by

$$D(r_{\Lambda_1}^{(m)}) = C^1(\Lambda_1^m) \otimes \mathcal{H}^{(n-m)}(\Lambda_2)$$

and

$$r_{\Lambda_1}^{(m)}((\psi\otimes\varphi)) = t_{0,\Lambda_1}^{(m)}(\psi)\ \|\varphi\|^2$$

for $\psi \in C^1(\Lambda_1^{(m)})$, $\varphi \in \mathcal{H}^{(n-m)}(\Lambda_2)$. But this latter form determines, by closure, the unique self-adjoint extension $(T_{0,\Lambda_1}^{(m)} \otimes 1_{\Lambda_2}^{(n-m)})^{**}$ of the tensor product of $T_{0,\Lambda_1}^{(m)}$ on $\mathcal{H}^{(m)}(\Lambda_1)$ and the identity $1_{\Lambda_2}^{(n-m)}$ on $\mathcal{H}^{(n-m)}(\Lambda_2)$.

However if $\psi \in D(T_{0,\Lambda_1}^{(m)})$, $\varphi \in \mathcal{H}^{(n-m)}(\Lambda_2)$ and $\chi \in D(\hat{s}_{\Lambda_1}^{(m)})$ then

$$\hat{s}_{\Lambda_1}^{(m)}(\chi, \psi\otimes\varphi) = (\chi\ ,\ T_{0,\Lambda_1}^{(m)}\psi \otimes \varphi)$$

as one checks by partial integration using the boundary condition

$$\frac{\partial\psi}{\partial n_x}(X_m) = 0 \qquad\qquad x \in X_m \cap \partial\Lambda_1$$

which is valid for all $\psi \in D(T_{0,\Lambda_1}^{(m)})$ (cf. 1.2.17.). Thus using part 3 of 1.2.6. one concludes that the self-adjoint operator $S^{(m)}_{\Lambda_1}$ associated with the closure of $\hat{S}^{(m)}_{\Lambda_1}$ is given by

$$S^{(m)}_{\Lambda_1} = \left(T^{(m)}_{0,\Lambda_1} \otimes 1^{(n-m)}_{\Lambda_2} \right)^{**} .$$

But we now have

$$t^{(n)}_{0,\Lambda_1 \cup \Lambda_2} \geqslant \sum_{m=0}^{n} \hat{S}^{(m)}_{\Lambda_1} + \hat{S}^{(n-m)}_{\Lambda_2}$$

in the sense of 1.2.3. and hence from part 2 of 1.2.15.

$$Q^{0}_{\Lambda_1 \cup \Lambda_2}(\beta, n) \leqslant \sum_{m=0}^{n} \mathrm{Tr}_{\mathcal{H}^{(m)}(\Lambda_1) \otimes \mathcal{H}^{(n-m)}(\Lambda_2)} \left(e^{-\beta (S^{(m)}_{\Lambda_1} + S^{(n-m)}_{\Lambda_2})} \right).$$

However using the foregoing identification of the $S^{(m)}_{\Lambda_1}$ etc. and the tensor product structure we conclude that

$$Q^{0}_{\Lambda_1 \cup \Lambda_2}(\beta, n) \leqslant \sum_{m=0}^{n} Q^{0}_{\Lambda_1}(\beta, m)\, Q^{0}_{\Lambda_2}(\beta, n-m) .$$

The inequality for Z_Λ then follows as in 2.1.5.

Finally we prove the increase property. Take Λ_1 and Λ_2 to be two parallelepipeds with the same centre and orientation and with $\Lambda_1 \subset \Lambda_2$. One can choose $\lambda^{(1)}, \dots, \lambda^{(\nu)} \geqslant 1$ such that if $x = (x^{(1)}, \dots, x^{(\nu)}) \in \Lambda_1$ then $\lambda x = (\lambda^{(1)} x^{(1)}, \dots, \lambda^{(\nu)} x^{(\nu)}) \in \Lambda_2$, and such that the image of $\bar{\Lambda}_1$ under this mapping is exactly $\bar{\Lambda}_2$. Introduce the notation $J = \lambda^{(1)} \lambda^{(2)} \dots \lambda^{(\nu)}$. Now if $\psi \in D(t^{(n)}_{0}, \Lambda_2)$ we define ψ_λ by

$$\psi_\lambda(x_1, \dots, x_n) = \psi(\lambda x_1, \dots, \lambda x_n)$$

and note that

$$\begin{aligned} t^{(n)}_{0,\Lambda_1}(\psi_\lambda) &= \frac{1}{n!} \int_{\Lambda_1^n} dx_1 \dots dx_n \sum_{i=1}^{n} |\nabla_{x_i} \psi(\lambda x_1, \dots, \lambda x_n)|^2 \\ &\geqslant J^{-n} \frac{1}{n!} \int_{\Lambda_2^{(n)}} dx_1 \dots dx_n \sum_{i=1}^{n} |\nabla_{x_i} \psi(x_1, \dots, x_n)|^2 \\ &= J^{-n}\, t^{(n)}_{0,\Lambda_2}(\psi) \end{aligned}$$

where the first step follows by changing variables and using $\lambda^{(i)} \geqslant 1$.

Further one finds straightforwardly that

$$\|\psi_\lambda\|^2 = J^{-n} \|\psi\|^2 .$$

Thus

$$t^{(n)}_{0,\Lambda_1}(\psi_\lambda) / \|\psi_\lambda\|^2 \geqslant t^{(n)}_{0,\Lambda_2}(\psi) / \|\psi\|^2 .$$

But it follows from this inequality and the minimax principle 1.2.14. for the eigenvalues $\lambda_m(T^{(n)}_{0,\Lambda})$ of $T^{(n)}_{0,\Lambda}$ that

$$\lambda_m(T^{(n)}_{0,\Lambda_1}) \geq \lambda_m(T^{(n)}_{0,\Lambda_2}) \qquad m = 1, 2, \ldots \; .$$

Hence

$$Q^{0}_{\Lambda_1}(\beta, n) \leq Q^{0}_{\Lambda_2}(\beta, n)$$

and

$$Z^{0}_{\Lambda_1}(\beta, \mu) \leq Z^{0}_{\Lambda_2}(\beta, \mu) \; .$$

2.1.7. The foregoing method to prove increase properties of the functions $\Lambda \to Q^{0}_{\Lambda}$, $\Lambda \to Z^{0}_{\Lambda}$ can be applied to other shaped regions than parallelepipeds but the result obtained is much weaker than that of 2.1.5. It is unclear whether the partition functions with elastic boundary conditions $(\sigma = 0)$ are in general increasing functions of Λ.

Proposition 2.1.8. Let Λ be a parallelepiped. The following limit exists

$$p^{\infty}(\beta, \mu) = \lim_{L^{(1)}, \ldots, L^{(\nu)} \to \infty} \frac{1}{V(\Lambda)} \log Z^{\infty}_{\Lambda}(\beta, \mu) \, , \quad \beta > 0, \; \mu < 0$$

and

$$p^{\infty}(\beta, \mu) = \sup_{L^{(1)}, \ldots, L^{(\nu)}} \frac{1}{V(\Lambda)} \log Z^{\infty}_{\Lambda}(\beta, \mu)$$

$p^{\infty}(\beta, \mu)$ is convex in β and μ and continuous in the pair (β, μ).

Proposition 2.1.9. The following limit exists

$$p^{0}(\beta, \mu) = \lim_{L^{(1)}, \ldots, L^{(\nu)} \to \infty} \frac{1}{V(\Lambda)} \log Z^{0}_{\Lambda}(\beta, \mu) \, , \quad \beta > 0, \; \mu < 0$$

and

$$p^{0}(\beta, \mu) = \inf_{L^{(1)}, \ldots, L^{(\nu)}} \frac{1}{V(\Lambda)} \log Z^{0}_{\Lambda}(\beta, \mu)$$

$p^{0}(\beta, \mu)$ is convex in β and μ and continuous in the pair (β, μ).

The functions $p^{0}(\beta, \mu)$ and $p^{\infty}(\beta, \mu)$ introduced by these propositions correspond to the thermodynamic pressure of the non-interacting system with purely elastic and repulsive boundary conditions respectively. We will prove below that they are in fact equal.

The proof of the existence of these functions follows from a standard argument based on the following three properties (we consider the elastic case

Invariance

$$\log Z^{\circ}_{\Lambda+a}(\beta,\mu) = \log Z^{\circ}_{\Lambda}(\beta,\mu) \qquad a \in R^{\nu}.$$

Boundedness

$$\frac{1}{V(\Lambda)} \log Z^{\circ}_{\Lambda}(\beta,\mu) \leq C .$$

Sub-additivity

$$\log Z^{\circ}_{\Lambda_1\cup\Lambda_2}(\beta,\mu) \leqslant \log Z^{\circ}_{\Lambda_1}(\beta,\mu) + \log Z^{\circ}_{\Lambda_2}(\beta,\mu) .$$

For the repulsive case one uses the lower boundedness and super-additivity. We will not repeat the details (cf. exercise 1) .

2.1.10. Next we will aim to prove that $P^{\infty}(\beta,\mu) = P^{\circ}(\beta,\mu)$. In fact in the non-interacting case under discussion this can be checked by explicit calculation but we will instead give an indirect proof which will be sufficiently general to accomodate positive decreasing interactions. First note that $t^{(n)}_{0,\Lambda} \leq t^{(n)}_{\infty,\Lambda}$ and hence by part 3 of 1.2.15. we have

$$Q^{\circ}_{\Lambda}(\beta,n) \geqslant Q^{\infty}_{\Lambda}(\beta,n)$$

and consequently

$$P^{\circ}(\beta,\mu) \geqslant P^{\infty}(\beta,\mu)$$

We need to prove this inequality in the opposite direction.

2.1.11 Let Λ_L be the parallelepiped

$$\Lambda_L = \left\{ x \in R^{\nu} ; -\frac{L^{(i)}}{2} \leqslant x^{(i)} \leqslant \frac{L^{(i)}}{2} , i = 1, 2, \cdots \right\} .$$

For each $b < \frac{L^{(i)}}{2}$, $i = 1, \ldots \nu$. we will define an isometry from $\mathcal{H}^{(n)}(\Lambda_L)$ to $\mathcal{H}^{(n)}(\Lambda_{L+2b})$. First we associate with each point $x \in \Lambda_{3L}$ a point $x_R \in \Lambda_L$ by noting that if $x \in \Lambda_{3L} \setminus \Lambda_L$ then there is a unique vector l_x with the two properties

1. $l_x = (n_x^{(1)} L^{(1)}, \ldots, n_x^{(\nu)} L^{(\nu)})$ with $n_x^{(i)} = 0$ or ± 1

2. $x + l_x \in \Lambda_L$

and we define then

$$x_R^{(i)} = - n_x^{(i)} L^{(i)} + (-1)^{n_x^{(i)}} x^{(i)} .$$

If $x \in \Lambda_L$ we set $x_R = x$.

Geometrically what we are doing is associating with each point $x_R \in \Lambda_L$ the set of points $x \in \Lambda_{3L}$ which may be obtained by reflection of x_R around a face, or faces, of Λ_L (cf. figure 8)

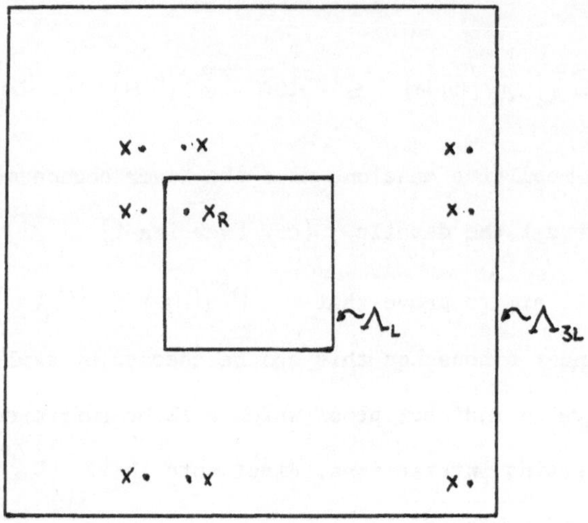

Fig. 8

In figure 8 each point x has the point x_R as image.

Next for each $\psi \in \mathcal{H}^{(n)}(\Lambda_L)$ we define $\psi^R \in \mathcal{H}^{(n)}(\Lambda_{3L})$ by

$$\psi^R(x_1, \ldots, x_n) = \psi(x_{1R}, \ldots, x_{nR}) .$$

Now introduce functions f_i, $i = 1, \ldots, \nu$ of one variable with the following properties

1. f_i is real, even, and once continuously differentiable.
2. $f_i(x) = 0$ for $|x| > L^{(i)}/2 + b$
3. $f_i(x) = 1$ for $|x| < L^{(i)}/2 - b$
4. $|f_i(x)|^2 + |f_i(-x - L^{(i)})|^2 = 1$ for $-L^{(i)}/2 - b \leq x \leq -L^{(i)}/2$
5. $\left| \frac{df_i(x)}{dx} \right|^2 \leq \frac{1}{b^2}$, $|f_i(x)|^2 \leq 1$.

Note that condition 4. can be written

$$\left[\left| f_i\left(y - \frac{L^{(i)}}{2}\right) \right|^2 - \frac{1}{2} \right] = - \left[\left| f_i\left(-y - \frac{L^{(i)}}{2}\right) \right|^2 - \frac{1}{2} \right] , \quad -b \leq y \leq b$$

i.e. $\quad Y \in [-b, b] \rightarrow \frac{1}{2} - \left| f_i \left(Y - \frac{L^{(i)}}{2} \right) \right|^2$ is antisymmetric.

Finally for $x \in R^\nu$ introduce f by

$$f(x) = \prod_{i=1}^{\nu} f_i(x^{(i)})$$

and define the mapping $V : \psi \in \mathcal{H}^{(n)}(\Lambda_L) \longrightarrow V\psi \in \mathcal{H}^{(n)}(\Lambda_{L+2b})$ by

$$(V\psi)(x_1, \ldots, x_n) = \psi^R(x_1, \ldots, x_n) \prod_{i=1}^{n} f(x_i) .$$

Lemma 2.1.12 :

The mapping V introduced above is an isometry, i.e.

$$(V\varphi, V\psi) = (\varphi, \psi)$$

for all $\varphi, \psi \in \mathcal{H}^{(n)}(\Lambda_L)$.

Introduce the notation $I = \Lambda_{L-2b}$, $S_I = \Lambda_L \setminus \Lambda_{L-2b}$ $S_E = \Lambda_{L+2b} \setminus \Lambda_L$ i.e. the sets represent the "interior", "inner surface" and "external surface" of Λ. Now

$$(V\varphi, V\psi) =$$

$$= \sum_{\substack{n_1\, n_2\, n_3 \\ n_1+n_2+n_3=n}} \int_{I^{n_1}} dX_{n_1} \int_{S_I^{n_2}} dY_{n_2} \int_{S_E^{n_3}} dZ_{n_3}\, \overline{\varphi^R(X_{n_1} \cup Y_{n_2} \cup Z_{n_3})}\, \psi^R(X_{n_1} \cup Y_{n_2} \cup Z_{n_3}) \, |f(Y_{n_2})|^2 \, |f(Z_{n_3})|^2$$

where

$$|f(Y)|^2 = \prod_{y \in Y} |f(y)|^2$$

and we have used

$$|f(x)|^2 = 1 \qquad x \in I .$$

Now if $Z_{n_3} = \{z_1, \ldots, z_{n_3}\}$ we denote by $Z_{n_3}^R$ the set of points $\{z_{1R}, \cdots, z_{n_3 R}\}$ obtained by reflecting the z_i with the map introduced above. Note that if $Z_{n_3} \subset S_E$ then $Z_{n_3}^R \subset S_I$. Thus

$$(V\varphi, V\psi) = \sum_{\substack{n_1\, n_2\, n_3 \\ n_1+n_2+n_3=n}} \int_{I^{n_1}} dX_{n_1} \int_{S_I^{n_2}} dY_{n_2} \int_{S_E^{n_3}} dZ_{n_3}\, \overline{\varphi(X_{n_1} \cup Y_{n_2} \cup Z_{n_3}^R)}\, \psi(X_{n_1} \cup Y_{n_2} \cup Z_{n_3}^R) \, |f(Y_{n_2})|^2 \, |f(Z_{n_3})|^2 .$$

Next by suitable reflections of coordinates the Z_{n_3} integrations can be replaced by integrations over $S_I^{n_3}$. Let us make these transformations and consider the value of the integrand for a fixed set $T_{n_2+n_3} = Y_{n_2} \cup Z_{n_3}^R$. There is a fixed factor $\overline{\varphi(X_{n_1} \cup T_{n_2+n_3})}\, \psi(X_{n_1} \cup T_{n_2+n_3})$ and various factors depending upon f

Let $t \in T_{n_2+n_3}$. If t is "near a face" of Λ_L e.g. if

$-L^{(i)}/2+b \leq t^{(i)} \leq L^{(i)}/2-b$ for $i=2,\dots,\nu$, then there are two terms which contribute factors. The first term arises when $t \in Y_{n_2}$ and has a t dependent factor $|f(t^{(1)})|^2$. The second term arises when $t \in Z^R_{n_3}$ and differs from the first only in the t dependent factor which becomes $|f(-t^{(1)}-L^{(1)})|^2$ or $|f(-t^{(1)}+L^{(1)})|^2$ depending whether $-L^{(1)}/2 \leq t^{(1)} \leq -L^{(1)}/2+b$ or $L^{(1)}/2-b \leq t^{(1)} \leq L^{(1)}/2$. In both cases the two factors add to give a t dependent factor which is exactly unity due to our choice of f. If on the other hand t is near an edge, e.g. $L^{(1)}/2-b \leq t^{(1)} \leq L^{(1)}/2, \dots, L^{(k)}/2-b \leq t^{(k)} \leq L^{(k)}/2$ and $-L^{(j)}/2+b \leq t^{(j)} \leq L^{(j)}/2-b$ for $j=k+1,\dots,\nu$ then there are 2^k different factors. One comes from $t \in Y_{n_2}$ and the remaining 2^k-1 come from the fact that $t \in Z^R_{n_3}$ can be reached by reflexion of 2^k-1 distinct points $z \in Z_{n_3}$. It is immediately checked however that the t-dependent factors of these terms add to give

$$\left(|f(t^{(1)})|^2+|f(-t^{(1)}+L^{(1)})|^2\right)\dots\left(|f(t^{(k)})|^2+|f(-t^{(k)}+L^{(k)})|^2\right)=1$$

Thus dealing successively with each $t \in T_{n_2+n_3}$ in this manner we are left with a factor unity and hence

$$(V\varphi, V\psi) = \sum_{\substack{n_1 n_2 n_3 \\ n_1+n_2+n_3=n}} \int_{I^{n_1}} dX_{n_1} \int_{S_I^{n_2}} dY_{n_2} \int_{S_I^{n_3}} dZ_{n_3} \overline{\varphi(X_{n_1}\cup Y_{n_2}\cup Z_{n_3})}\,\psi(X_{n_1}\cup Y_{n_2}\cup Z_{n_3})$$

$$= \int_{I^n} dX_n \overline{\varphi(X_n)}\,\psi(X_n) = (\varphi,\psi).$$

The motive for introducing the above isometry is the estimate which follows.

<u>Lemma 2.1.13.</u>

<u>Take $\psi \in D(t^{(n)}_{0,\Lambda_L})$. It follows that $V\psi \in D(t^{(n)}_{\infty,\Lambda_{L+2b}})$</u>

<u>and</u>

$$t^{(n)}_{\infty,\Lambda_{L+2b}}(V\psi) \leq t^{(n)}_{0,\Lambda_L}(\psi) + 2n\nu\|\psi\|^2/b^2.$$

First assume that $\psi \in C^1(\Lambda^n_L)$ and satisfies the condition $\partial\psi/\partial n_x = 0$ for $x \in \partial\Lambda$ it then follows that $\psi^R \in C^1(\Lambda^n_{3L})$. But the first two properties of the functions then guarantee that $V\psi$ is in the stated domain. Next let us calculate

$$t^{(n)}_{\infty,\Lambda_{L+2b}}(V\psi) = \int_{\Lambda^n_{L+2b}} dX_n \sum_{x\in X_n} \left| (\underline{\nabla}_x \psi^R(X_n)) f(X_n) + \psi^R(X_n) \underline{\nabla}_x f(X_n) \right|^2 .$$

If we explicitly take the square in the integrand we find three terms which we discuss separately. First we have

$$\int_{\Lambda^n_{L+2b}} dX_n \sum_{x\in X_n} | \underline{\nabla}_x \psi^R(X_n)|^2 \, |f(X_n)|^2 = t^{(n)}_{0,\Lambda_L}(\psi)$$

where the equality follows by the argument used in the proof of 2.1.12. Secondly we find

$$\int_{\Lambda^n_{L+2b}} dX_n \, | \psi^R(X_n)|^2 \sum_{x\in X_n} | \underline{\nabla}_x f(X_n)|^2$$

which is a sum of $n\nu$ terms the first of which is

$$\int_{\Lambda^n_{L+2b}} dX_n \, | \psi^R(X_n)|^2 \, | \partial/\partial x^{(1)}_1 f_1(x^{(1)}_1)|^2 \cdots |f_\nu(x^{(\nu)}_1)|^2 \, |f(x_2)|^2 \cdots |f(x_n)|^2 .$$

Again using the arguments of 2.1.12. we can change the integration region, eliminate the $|f|^2$ and identify this term with

$$\int_{\Lambda^n_L} dX_n \, | \psi(X_n)|^2 \left(\left| \frac{\partial}{\partial x^{(1)}_1} f(x^{(1)}_1) \right|^2 + \left| \frac{\partial}{\partial x^{(1)}_1} f(x^{(1)}_{1R}) \right|^2 \right) \leq \frac{2}{b^2} \|\psi\|^2$$

where the bound follows from property 5 of the f_i. Thus the second term is majorized by $2n\nu\|\psi\|^2/b^2$. The third term is given by

$$\frac{1}{2} \int_{\Lambda^n_{L+2b}} dX_n \sum_{x\in X_n} \underline{\nabla}_x | \psi^R(X_n)|^2 . \, \underline{\nabla}_x |f(X_n)|^2$$

where we have used the reality of the f_i. Again there are $n\nu$ terms each of which is an integral of a function of the type

$$\int_{-L^{(i)}-2b}^{L^{(i)}+2b} dx^{(i)}_j \left(\frac{\partial}{\partial x^{(i)}_j} | \psi^R(X_n)|^2 \right) \left(\frac{\partial}{\partial x^{(i)}_j} |f(X_n)|^2 \right) .$$

But the second factor vanishes for $-L^{(i)}+b \leq x^{(i)}_j \leq L^{(i)}-b$ and hence the integral is only over the two end intervals. However both these integrals vanish for the following reason. Consider the integral over $(L^{(i)}-b, \; L^{(i)}+b)$ and change variables to $Y = L^{(i)} - x^{(i)}_j$ to transform the integral to the interval $(-b, b)$. Now as a result of property 4 of the f_i the second factor is an even function of Y and the first factor is odd because of the extension by reflection. Thus the integral is zero and similarly the complete third term must vanish.

Collecting the estimates of the three individual terms we arrive at the stated result for the subclass of ψ considered. But this set is a core of $t^{(n)}_{0\Lambda_L}$ (cf. 1.1.15. and the subsequent definitions) and hence the result then follows by closure.

Proposition 2.1.14. For all $\beta > 0$

$$Q^{0}_{\Lambda_L}(\beta, n)\, e^{-2n\nu\beta/b^2} \leq Q^{\infty}_{\Lambda_{L+2b}}(\beta, n) \leq Q^{0}_{\Lambda_{L+2b}}(\beta, n)$$

and if $\mu < 0$ then

$$Z^{0}_{\Lambda_L}(\beta, \mu - 2\nu/b^2) \leq Z^{\infty}_{\Lambda_{L+2b}}(\beta, \mu) \leq Z^{0}_{\Lambda_{L+2b}}(\beta, \mu)$$

As a consequence of these estimates and the existence of the limit (and its continuity (2.1.9.)

$$P^{0}(\beta, \mu) = \lim_{L^{(1)}, \ldots, L^{(\nu)} \to \infty} \frac{1}{V(\Lambda)} \log Z^{0}_{\Lambda}(\beta, \mu) \qquad , \mu < 0 , \beta > 0$$

it follows that each of the subsequent limits exists

$$P^{\infty}(\beta, \mu) = \lim_{L^{(1)}, \ldots, L^{(\nu)} \to \infty} \frac{1}{V(\Lambda)} \log Z^{\infty}_{\Lambda}(\beta, \mu) \qquad , \mu < 0 , \beta > 0$$

$$P^{\sigma}(\beta, \mu) = \lim_{L^{(1)}, \ldots, L^{(\nu)} \to \infty} \frac{1}{V(\Lambda)} \log Z^{\sigma}_{\Lambda}(\beta, \mu) \qquad \sigma \geq 0 , \mu < 0 , \beta > 0$$

and moreover,

$$P^{\infty}(\beta, \mu) = P^{\sigma}(\beta, \mu) = P^{0}(\beta, \mu) .$$

Note that this proposition gives a proof of the existence of $P^{\infty}(\beta, \mu)$ which is completely independent of the estimate of 2.1.5. ; the latter is replaced by 2.1.9. and 2.1.13. and the consequent inequalities stated at the beginning of the proposition. This remark will be of importance in section §2 where we consider interacting particles because with suitable assumptions we can establish the analogoues of 2.1.9. and 2.1.13. but not the estimate 2.1.5.

Secondly we emphasize that the proposition deals with the case $\sigma \geq 0$ only, i.e. repulsive walls, we consider $\sigma < 0$ below.

To establish the first estimate note again that the set D of $\psi \in D(t^{(n)}_{0,\Lambda})$ with $\partial\psi/\partial n_x = 0$ for $x \in \partial\Lambda$ is a core of $t^{(n)}_{0,\Lambda}$

Let us use 1.2.15. and 2.1.13. We have

$$Q^{0}_{\Lambda_L}(\beta, n) = \sup_{\mathcal{F} \subset D} \sum_{\psi \in \mathcal{F}} \exp\{-\beta t^{(n)}_{0,\Lambda_L}(\psi)\}$$

$$\leq \sup_{\mathcal{F} \subset D} \sum_{\psi \in \mathcal{F}} \exp\left\{-\beta t^{(n)}_{\infty, \Lambda_{L+2b}}(V\psi)\right\} e^{2\beta n\nu/b^2}$$

$$\leq Q^{\infty}_{\Lambda_{L+2b}}(\beta, n)\, e^{2\beta n\nu/b^2}$$

where the last step uses that V is an isometry and 1.2.15. The second estimate for the Q has already been mentioned in 2.1.10. ; the estimates for the Z follow immediately from definition.

Next using the existence of the limit of 2.1.9. we find

$$P^{\circ}(\beta, \mu - 2\nu/b^2) \leq \hat{P}^{\infty}(\beta,\mu) \leq P^{\circ}(\beta,\mu)$$

where $\hat{P}^{\infty}$ is any limit point of $V(\Lambda_L)^{-1} \log Z^{\infty}_{\Lambda_L}(\beta,\mu)$ as $L^{(1)}, \ldots, L^{(\nu)} \to \infty$ However after the limit $L^{(1)}, \ldots, L^{(\nu)} \to \infty$ there are no longer any restrictions on the size of b so we can take the limit $b \to \infty$ and use the continuity of P° as a function of μ to conclude that

$$P^{\circ}(\beta,\mu) = \hat{P}^{\infty}(\beta,\mu) .$$

But as this is true for an arbitrary limit point $\hat{P}^{\infty}$ it is true for all limit points. Thus $P^{\infty}(\beta,\mu)$ exists and

$$P^{\circ}(\beta,\mu) = P^{\infty}(\beta,\mu) .$$

Finally if $\sigma \geq 0$ then we have

$$t^{(n)}_{\infty,\Lambda} \geq t^{(n)}_{\sigma,\Lambda} \geq t^{(n)}_{0,\Lambda}$$

which can be translated, with the aid of 1.2.15., into the inequality

$$Z^{\infty}_{\Lambda}(\beta,\mu) \leq Z^{\sigma}_{\Lambda}(\beta,\mu) \leq Z^{\circ}_{\Lambda}(\beta,\mu) .$$

The existence of $P^{\sigma}(\beta,\mu)$ and its identification with $P^{\circ}(\beta,\mu)$ is then a direct consequence of the last inequality and the foregoing result of the proposition.

2.1.15. Finally we will consider the case $\sigma < 0$. Note that

$$t^{(n)}_{0,\Lambda} \geq t^{(n)}_{\sigma,\Lambda} \qquad , \sigma < 0$$

and hence, from 1.2.15, we have

$$Q^{\circ}_{\Lambda}(\beta,n) \leq Q^{\sigma}_{\Lambda}(\beta,n) \qquad , \sigma < 0$$

and

$$Z^{\circ}_{\Lambda}(\beta,\mu) \leq Z^{\sigma}_{\Lambda}(\beta,\mu) \qquad , \sigma < 0 .$$

Thus we must establish an estimate in the other direction if we want to quantitatively compare the different partition functions. This will be done by utilisation of an argument similar to that developed in 2.1.6.

Proposition 2.1.16. Let Λ_1 and Λ_2 be open bounded and connected sets in R^ν with smooth boundaries such that $\Lambda_2 \subset \Lambda_1$ and $\partial\Lambda_1 \cap \partial\Lambda_2 = \phi$. Let $\Lambda_s = \Lambda_1 \setminus \Lambda_2$. Let $\hat{Q}^{\sigma}_{\Lambda_s}(\beta, n)$ denote the partition function defined with a Hamiltonian $T^{(n)}_{\sigma,\Lambda_s}$ where $\sigma = 0$ on $\partial\Lambda_2$ and $\sigma < 0$ on $\partial\Lambda_1$. It follows that

$$Q^{\sigma}_{\Lambda_1}(\beta, n) \leq \sum_{m=0}^{n} Q^{0}_{\Lambda_2}(\beta, n)\, \hat{Q}^{\sigma}_{\Lambda_s}(\beta, n-m).$$

Consequently for $b < L = \min L^{(i)}$ and $\mu < -\nu|\sigma|(|\sigma| + 2/b)$ there is a $C > 0$ such that

$$Z^{\sigma}_{\Lambda_{L+2b}}(\beta, \mu) \leq Z^{0}_{\Lambda_L}(\beta, \mu) \exp\{ b V(\Lambda_L) C/L \}.$$

Note that in the rather terse statement concerning the definition of $\hat{Q}^{\sigma}_{\Lambda_s}$ we have attempted to indicate that we are returning to the general definition of $t^{(n)}_{\sigma,\Lambda_s}$ given in 1.3.4. with σ a function over the surface of Λ_s which is no longer constant but has the stated properties.

The proof of the first part of the proposition is almost identical to that of 2.1.6.

First one makes a decomposition

$$t^{(n)}_{\sigma,\Lambda_1}(\psi) = t^{(n)}_{\sigma,\Lambda_2 \cup \Lambda_s}(\psi) = \sum_{m=0}^{n} s^{(m)}_{0,\Lambda_2}(\psi) + \hat{s}^{(n-m)}_{\sigma,\Lambda_s}(\psi)$$

where

$$s^{(m)}_{0,\Lambda_2}(\psi) = \int_{\Lambda_2} dX_m \int_{\Lambda_s} dY_{n-m} \sum_{x \in X_m} |\nabla_x \psi(X_m \cup Y_{n-m})|^2$$

$$\hat{s}^{(n-m)}_{\sigma,\Lambda_s}(\psi) = \int_{\Lambda_2} dX_m \int_{\Lambda_s} dY_{n-m} \sum_{y \in Y_{n-m}} |\nabla_y \psi(X_m \cup Y_{n-m})|^2 + \int_{\Lambda_2} dX_m \int_{\partial\Lambda_s} dS_{n-m}\, \sigma(Y) |\psi(X \cup Y)|^2$$

and $\sigma(Y) = \sigma$ on $\partial\Lambda_1^{n-m}$ and equals zero on $\partial\Lambda_2^{n-m}$. But then the arguments of 2.1.6. can be used to deduce that $s^{(m)}_{0,\Lambda_2}$ is a restriction of the form determining $(T^{(m)}_{0,\Lambda_2} \otimes \mathbb{1}^{(n-m)}_{\Lambda_s})^{**}$ and similarly $\hat{s}^{(n-m)}_{0,\Lambda_s}$ is associated with an operator $(\mathbb{1}^{(m)}_{\Lambda_2} \otimes \hat{T}^{(n-m)}_{\sigma,\Lambda_s})^{**}$ where $\hat{T}^{(n-m)}_{\sigma,\Lambda_s}$ is the kinetic energy operator with mixed boundary conditions ($\sigma = 0$ on $\partial\Lambda_2$, $\sigma < 0$ on $\partial\Lambda_1$)

The first inequality then follows as 2.1.6.

By summation one immediately deduces that

$$Z^{\sigma}_{\Lambda_{L+2b}}(\beta,\mu) \le Z^{0}_{\Lambda_L}(\beta,\mu)\, \hat{Z}^{\sigma}_{\Lambda_S}(\beta,\mu)$$

where $\Lambda_S = \Lambda_{L+2b} \setminus \Lambda_L$ and $\hat{Z}^{\sigma}_{\Lambda_S}$ is defined with mixed boundary conditions. It remains to show that $\hat{Z}^{\sigma}_{\Lambda_S}$ has an appropriate upper bound. This would follow if we could apply the calculations of 2.1.2. but these are of course only valid for parallelepipeds. This difficulty can however be circumvented by decomposing Λ_S into 2ν parallelepipeds Λ_i adjacent to the 2ν faces of Λ_L and applying the above reasoning once again to deduce a sub-multiplicative bound of the form

$$\hat{Z}^{\sigma}_{\Lambda_S}(\beta,\mu) \le \prod_{i=1}^{2\nu} \hat{Z}^{\sigma}_{\Lambda_i}(\beta,\mu) .$$

The partition functions are again defined with mixed boundary conditions ; see figure 9 where we indicate the decomposition and the boundary conditions .

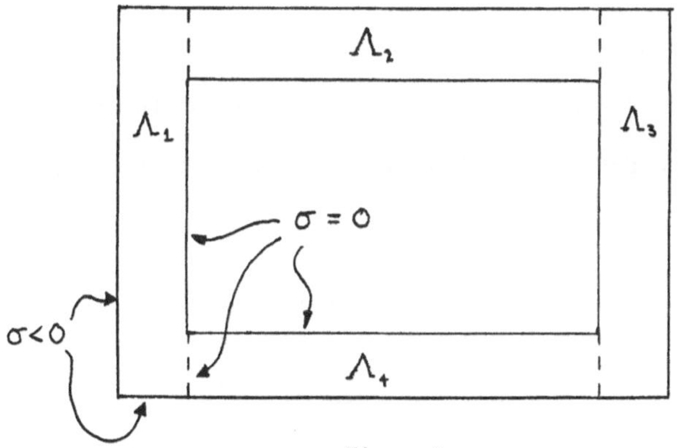

Fig. 9

But as $\sigma < 0$ we can use the decrease of $\sigma \to t^{(n)}_{\sigma,\Lambda}$ and 1.2.15. to deduce that

$$\prod_{i=1}^{2\nu} \hat{Z}^{\sigma}_{\Lambda_i}(\beta,\mu) \le \prod_{i=1}^{2\nu} Z^{\sigma}_{\Lambda_i}(\beta,\mu) .$$

Combining these two bounds and using 2.1.2. we arrive at the desired result.

<u>Proposition 2.1.17.</u> <u>As a consequence of 2.1.16. and the existence of the limit (2.1.9.)</u>

$$P^{0}(\beta,\mu) = \lim_{L^{(1)},\dots,L^{(\nu)}\to\infty} \frac{1}{V(\Lambda)} \log Z^{0}_{\Lambda}(\beta,\mu) \quad , \beta>0, \mu<0 .$$

<u>It follows that if $\beta > 0$, $\mu < -\nu\sigma^2$ and $\sigma < 0$ then the following limit exists</u>

$$p^{\sigma}(\beta,\mu) = \lim_{L^{(1)},\dots,L^{(\nu)}\to\infty} \frac{1}{V(\Lambda)} \log Z^{\sigma}_{\Lambda}(\beta,\mu)$$

<u>and moreover,</u>

$$p^{\sigma}(\beta,\mu) = p^{\circ}(\beta,\mu) .$$

The proof follows immediately form 2.1.15. and 2.1.16. because we have

$$Z^{\circ}_{\Lambda_{L+2b}}(\beta,\mu) \leq Z^{\sigma}_{\Lambda_{L+2b}}(\beta,\mu) \leq Z^{\circ}_{\Lambda_L}(\beta,\mu)\exp\{bV(\Lambda_L)c/L\}$$

whenever $\beta>0$ and $\mu < -\nu|\sigma|(|\sigma|+2|b)$ where $b<L$. As L is increased b can also be increased but we remain with the restriction $\mu < -\nu\sigma^2$.

<u>2.1.18.</u> At this point we have completed the proof that the thermodynamic pressure of a non-interacting system in the grand canonical ensemble is independent of the boundary conditions used in its definition if these boundary conditions are of the elastic type and the chemical potential is suitably restricted. We have only discussed explicitly the case where the elasticity is constant on the surface of the finite regions entering in the definitions but this restriction can be immediately lifted in the following manner.

If $t^{(n)}_{\sigma,\Lambda}$ is defined with σ varying on the surface of Λ but with $\sigma(x_m) \geq \sigma_m$ then

$$t^{(n)}_{\infty,\Lambda} \geq t^{(n)}_{\sigma,\Lambda} \geq t^{(n)}_{\sigma_m,\Lambda}$$

where $t^{(n)}_{\sigma_m,\Lambda}$ is defined with the constant elasticity σ_m. But then with a similar notation and 1.2.15. we deduce that

$$Z^{\infty}_{\Lambda}(\beta,\mu) \leq Z^{\sigma}_{\Lambda}(\beta,\mu) \leq Z^{\sigma_m}_{\Lambda}(\beta,\mu) .$$

However with appropriate restrictions on μ the thermodynamic pressures defined by using the partition function at the right and left of this inequality are the same and thus the intermediate function also gives the same result.

One can also deal in a similar manner with periodic boundary conditions. For example the form $t^{(n)}_{\Lambda}(\theta;\cdot)$ appropriate to the description of the kinetic energy operator with periodic boundary conditions has as domain a sub-domain of $D(t^{(n)}_{0,\Lambda})$ specified by conditions of the schematic form

$$\psi(0) = e^{i\theta}\psi(L)$$

and one has

$$t^{(n)}_{\infty,\Lambda} \geqslant t^{(n)}_{\Lambda}(\theta;\cdot) \geqslant t^{(n)}_{0,\Lambda}$$

(cf. 1.2.17.). Hence the partition function defined with periodic boundary conditions takes values between those of $Z^{\infty}_{\Lambda}(\beta,\mu)$ and $Z^{0}_{\Lambda}(\beta,\mu)$. Thus the thermodynamic pressure defined with these boundary conditions is equal to $P^{\infty}(\beta,\mu)$.

In the next section we reconsider the behaviour of the thermodynamic pressure for particles interacting with a repulsive interaction.

§2 - THE INTERACTING SYSTEM

2.2.1. A completely satisfactory discussion of the thermodynamic pressure of interacting particles should certainly cover an inter- particle interaction which is repulsive at short distances and possibly attractive at larger distances ; these latter properties being characteristic of the behaviour generally encountered in nature. We will however only consider positive, i.e. completely repulsive, interactions ; this class of interactions can be considered as an extreme case, a case which emphasizes the physical repulsion at short distances. An opposite extreme would be to consider interactions which are negative, i.e. completly attractive, at large distances. Let us briefly review the qualitative characteristics of these two extreme cases in the classical limit, i.e. ignoring the kinetic energy of the particles.

If $U(X_n)$ denotes the energy of interaction of n particles in the configuation X_n then the classical canonical partition function C_Λ is defined by

$$C_\Lambda(\beta,n) = \int_\Lambda dX_n\, e^{-\beta U(X_n)}.$$

If the interaction is completely repulsive then

$$U(X_n \cup Y_m) - U(X_n) - U(Y_m) \geqslant 0$$

for all possible configurations. It follows easily that if $\Lambda_1 \cap \Lambda_2 = \phi$ then

$$C_{\Lambda_1 \cup \Lambda_2}(\beta, n) \leq \sum_{m=0}^{n} C_{\Lambda_1}(\beta, m)\, C_{\Lambda_2}(\beta, n-m) \;. \qquad (*)$$

If however the interaction is completely attractive at distances larger than R then

$$U(X_n \cup Y_m) - U(X_n) - U(Y_m) \leq 0$$

for all configurations X_n, Y_m such that $d(X_n, Y_m) = \min_{x \in X_n} \min_{y \in Y_m} |x-y| > R$ it then follows that if $d(\Lambda_1, \Lambda_2) > R$ then

$$C_{\Lambda_1 \cup \Lambda_2}(\beta, n) \geq \sum_{m=0}^{n} C_{\Lambda_1}(\beta, m)\, C_{\Lambda_2}(\beta, n-m) \;. \qquad (**)$$

Note that $(*)$ and $(**)$ are in opposite directions ; $(*)$ is analogous to the result obtained in 2.1.6 for the non-interacting system with elastic boundary conditions $\sigma = 0$ whilst $(**)$ is analogue to the result of 2.1.5. obtained with infinitely repulsive boundary conditions.

These remarks indicate that it is natural to base the discussion of particles with a repulsive interaction on the elastic boundary conditions but to combine long range attraction with repulsive boundary conditions. The latter choice is the starting point of Fisher and Ruelle's discussion of the thermodynamic pressure ; we will follow the former course.

Let us now give in detail the assumptions we will need the interactions to satisfy.

2.2.2. We define a positive interaction to be a function U from the finite sets $X_n \subset R^\nu$ to $U(X_n) \in [0, \infty]$ which is sufficiently integrable to ensure that the sets

$$D(u_\Lambda^{(n)}) = \left\{ \psi \;;\; \psi \in \mathcal{H}^{(n)}(\Lambda) \,,\; \int_\Lambda dX_n\, U(X_n)\, |\psi(X_n)|^2 < \infty \right\}$$

are dense in $\mathcal{H}^{(n)}(\Lambda)$ for each n and $\Lambda \subset R^\nu$.

Actually this last restriction is stronger than necessary. In particular it excludes particles with hard cores. As the inclusion of hard cores entails the additional discussion of the boundary conditions imposed upon the cores we have made the foregoing restriction to avoid irrelevant complication. In fact the discussion of hard core particles is easier than the discussion of point particles and more general interactions can be accomodated. We refer the reader to the bibliography at the end of the chapter.

We associate with each U a family of forms

$$u^{(n)}_{\Lambda}(\psi) = \int_{\Lambda} dX_n \, U(X_n)\, |\psi(X_n)|^2 \qquad , \qquad \psi \in D(u^{(n)}_{\Lambda})$$

and note that these forms are positive, densely defined, and closed (the last property can, for example, be deduced from 1.2.7.) We denote by $U^{(n)}_{\Lambda}$ the positive self-adjoint operators, on $\mathcal{H}^{(n)}(\Lambda)$, associated with these forms by 1.2.6.

We will now impose various conditions on U. The first property that we demand is that $D(u^{(n)}_{\Lambda})$ should have a dense intersection with the domains of the closures of the forms $t^{(n)}_{\infty,\Lambda}$ and $t^{(n)}_{\sigma,\Lambda}$. With these conditions we can introduce the forms

$$h^{(n)}_{\sigma,\Lambda}(\psi) = \tilde{t}^{(n)}_{\sigma,\Lambda}(\psi) + u^{(n)}_{\Lambda}(\psi) \qquad , \qquad \psi \in D(\tilde{t}^{(n)}_{\sigma,\Lambda}) \cap D(u^{(n)}_{\Lambda})$$

$$h^{(n)}_{\infty,\Lambda}(\psi) = \tilde{t}^{(n)}_{\infty,\Lambda}(\psi) + u^{(n)}_{\Lambda}(\psi) \qquad , \qquad \psi \in D(\tilde{t}^{(n)}_{\infty,\Lambda}) \cap D(u^{(n)}_{\Lambda})$$

and be assured that these forms are densely defined, lower semi-bounded, and closable We denote by $H^{(n)}_{\sigma,\Lambda}$ and $H^{(n)}_{\infty,\Lambda}$ the self-adjoint operators associated with their closures. These operators are generalized sums $T^{(n)}_{\sigma,\Lambda} \dotplus U^{(n)}_{\Lambda}$ etc. of the kinetic energy and interaction operators and correspond to total Hamiltonians with elastic and repulsive boundary conditions respectively ; they are self-adjoint differential operators

$$\psi(X_n) \longrightarrow - \sum_{x \in X_n} \nabla_x^2 \psi(X_n) + U(X_n)\psi(X_n)$$

with the boundary conditions

$$\partial\psi / \partial n_x = \sigma \psi(X_n) \qquad\qquad x \in X_n \cap \partial\Lambda$$

and

$$\psi(X_n) = 0 \qquad\qquad X_n \cap \partial\Lambda \neq \emptyset$$

respectively.

2.2.3. Concerning the behaviour of the interaction as a function over the finite sets $X \subset R^{\nu}$ we also assume

1. $U(X_n + a) = U(X_n)$, $a \in R^{\nu}$, $X_n \subset R^{\nu}$
2. $U(X_n \cup Y_m) \geq U(X_n) + U(Y_m)$, $X_n \subset R^{\nu}$, $Y_m \subset R^{\nu}$
3. U is decreasing in the following sense.

Let C be a mapping of R^{ν} into R^{ν} whose action is denoted by

$$x \in R^\nu \longrightarrow x_c \in R^\nu \qquad \text{and}$$

$$X_n = \{x_1, \ldots, x_n\} \subset R^\nu \longrightarrow X_n^c = \{x_{1c}, \ldots, x_{nc}\} \subset R^\nu .$$

The mapping C is said to be a contraction if

$$|x-y| \geq |x_c - y_c| \quad , \quad x, y \in R^\nu$$

and if

$$|x-y| > |x_c - y_c|$$

for at least one pair $x, y \in R^\nu$.

The interaction is said to be decreasing if for each contraction C

$$U(X_n) \leq U(X_n^c) .$$

Note that by condition 1 $U(\{x\})$ is a constant ; we adopt the convention that it is zero.

2.2.4. Let us now repeat the definitions of the partition functions given in 2.1.1. but replace $T_{\sigma,\Lambda}^{(n)}$ by $H_{\sigma,\Lambda}^{(n)}$ etc. e.g. we now set

$$Q_\Lambda^\sigma(\beta, n) = Tr_{\mathcal{H}^{(n)}(\Lambda)}\left(e^{-\beta H_{\sigma,\Lambda}^{(n)}}\right) \qquad , \beta > 0$$

etc. We will retain the same notation Q_Λ, Z_Λ and list the results of §1 which are unchanged by the new definitions indicating at each step the required changes in the proofs.

Proposition 2.2.5. The estimates of proposition 2.1.2. remain valid for the redefined partition functions.

It follows immediately from $U_\Lambda^{(n)} \geq 0$ and 1.2.15. that

$$Tr_{\mathcal{H}^{(n)}(\Lambda)}\left(e^{-\beta H_{0,\Lambda}^{(n)}}\right) \leq Tr_{\mathcal{H}^{(n)}(\Lambda)}\left(e^{-\beta T_{0,\Lambda}^{(n)}}\right) .$$

Proposition 2.2.6. The convexity and continuity properties of proposition 2.1.3. remain valid .

The proof is identical.

Proposition 2.2.7. The results of proposition 2.1.4. remain valid.

The convergence property is deduced by noting that $H_{\sigma,\Lambda}^{(n)}$ has discrete spectrum with finite multiplicity because $T_{\sigma,\Lambda}^{(n)}$ has this property and $U_\Lambda^{(n)} \geq 0$ (apply 1.2.14.). Further $h_{\sigma,\Lambda}^{(n)} \rightarrow h_{\infty,\Lambda}^{(n)}$ as $\sigma \rightarrow \infty$ in the sense of 1.2.9. and thus the individual eigenvalues of $H_{\sigma,\Lambda}^{(n)}$ converge to those of $H_{\infty,\Lambda}^{(n)}$ by 1.2.10.

Proposition 2.2.8. The subadditivity and increase properties of $Q^o_\Lambda(\beta, n)$ and $Z^o_\Lambda(\beta, n)$ given in proposition 2.1.6. remain valid.

The proof of the sub-additivity property similar to that of 2.1.6. but uses the additional property

$$U(X_n \cup Y_m) \geq U(X_n) + U(Y_m)$$

of 2.2.3. One uses this property to show that $h^{(n)}_{0,\Lambda_1\cup\Lambda_2}$ is greater than a sum of forms of the type

$$S^{(m)}_{0,\Lambda_1}(\psi) = \int_{\Lambda_1} dX_m \int_{\Lambda_2} dY_{n-m} \left\{ \sum_{x \in X_n} |\nabla_x \psi(X_m \cup Y_{n-m})|^2 + U(X_m)|\psi(X_m \cup Y_{n-m})|^2 \right\}$$

where $\psi \in D(h^{(n)}_{0,\Lambda_1\cup\Lambda_2})$ and then one demonstrates that this form is a restriction of the form determining the operator $(H^{(m)}_{0,\Lambda_1} \otimes 1\!\!1^{(n-m)}_{\Lambda_2})^{**}$. The latter step is achieved by extending $S^{(m)}_{0,\Lambda_1}$ to $\hat{S}^{(m)}_{0,\Lambda_1}$ where $D(\hat{S}^{(m)}_{0,\Lambda_1})$ is taken to be the union of $D(h^{(n)}_{0,\Lambda_1\cup\Lambda_2})$ and the domain of the form determining $(H^{(m)}_{0,\Lambda_1} \otimes 1\!\!1^{(n-m)}_{\Lambda_2})^{**}$. But $\hat{S}^{(m)}_{0,\Lambda_1}$ is then closable. The first term, the kinetic energy, is closable by 2.1.6. and it is easily argued that the second term, the interaction energy, is closable and its closure has as domain all $\psi \in \mathcal{H}^{(n)}(\Lambda_1 \cup \Lambda_2)$ which are square integrable with respect to $dX_m\, dY_{n-m}\, U(X_m)$. Thus the sum is closable by 1.2.9. Now one can follow the argument of 2.1.6. and apply property 3 of 1.2.6. to deduce that $\hat{S}^{(m)}_{0,\Lambda_1}$ determines the appropriate operator. In this manner one finds that

$$H^{(n)}_{0,\Lambda_1\cup\Lambda_2} \geq \bigoplus_{m=0}^{n} \left[(H^{(m)}_{0,\Lambda_1} \otimes 1\!\!1^{(n-m)}_{\Lambda_2})^{**} \dot{+} (1\!\!1^{(m)}_{\Lambda_1} \otimes H^{(n-m)}_{0,\Lambda_2})^{**} \right]$$

and the desired result again follows from 1.2.15.

To deduce the increase property we use the notation of 2.1.6. and remark that

$$u^{(n)}_{0,\Lambda_1}(\psi_\lambda) = \int_{\Lambda_1} dX_n\, U(X_n)\, |\psi(\lambda X_n)|^2$$

$$= J^{-n} \int_{\Lambda_2} dX_n\, U(\lambda^{-1}X_n)\, |\psi(X_n)|^2$$

but

$$X_n = \{x_1, \ldots, x_n\} \longrightarrow \lambda^{-1}X_n = \{\lambda^{-1}x_1, \ldots, \lambda^{-1}x_n\}$$

is a contraction. So

$$u^{(n)}_{\Lambda_1}(\psi_\lambda)/\|\psi_\lambda\|^2 \geq u^{(n)}_{\Lambda_2}(\psi)/\|\psi\|^2 .$$

Combining this estimate with that of 2.1.6. we find

$$h^{(n)}_{0,\Lambda_2}(\psi_\lambda)/\|\psi_\lambda\|^2 \geq h^{(n)}_{0,\Lambda_2}(\psi)/\|\psi\|^2$$

and the rest of the proof is unchanged.

Proposition 2.2.9. The statement of proposition 2.1.9. remains valid thus the following limit exists

$$P^0(\beta,\mu) = \lim_{L^{(1)},\dots,L^{(\nu)}\to\infty} \frac{1}{V(\Lambda)} \log Z^0_\Lambda(\beta,\mu) \quad , \beta>0, \mu<0$$

and

$$P^0(\beta,\mu) = \inf_{L^{(1)},\dots,L^{(\nu)}} \frac{1}{V(\Lambda)} \log Z^0_\Lambda(\beta,\mu) \, .$$

The proof relies upon the boundedness property of 2.2.5. and the subadditivity of 2.2.8.

Proposition 2.2.10. The estimates of rproposition 2.1.14. remain valid and consequently the subsequent limits exist

$$P^\infty(\beta,\mu) = \lim_{L^{(1)},\dots,L^{(\nu)}\to\infty} \frac{1}{V(\Lambda)} \log Z^\infty_\Lambda(\beta,\mu) \quad , \beta>0, \mu<0$$

$$P^\sigma(\beta,\mu) = \lim_{L^{(1)},\dots,L^{(\nu)}\to\infty} \frac{1}{V(\Lambda)} \log Z^\sigma_\Lambda(\beta,\mu) \quad , \sigma\geq 0, \beta>0, \mu<0$$

and moreover,

$$P^\infty(\beta,\mu) = P^\sigma(\beta,\mu) = P^0(\beta,\mu) \, .$$

To establish this result we need to supplement the estimate of lemma 2.1.13. by the estimate

$$u^{(n)}_{\Lambda_{L+2b}}(V\psi) \leq u^{(n)}_{\Lambda_L}(\psi) \, .$$

To prove this we note that

$$u^{(n)}_{\Lambda_{L+2b}}(V\psi) = \int_{\Lambda_{L+2b}} dX_n \, U(X_n) \, |(V\psi)(X_n)|^2$$

$$\leq \int_{\Lambda_{L+2b}} dX_n \, U(X^R_n) |(V\psi)(X_n)|^2$$

because the mapping $X_n \to X^R_n$ is a contraction. Repeating the proof of lemma 2.1.12. applied to the right hand side of this inequality we find

$$\int_{\Lambda_{L+2b}} dX_n \, U(X^R_n) |(V\psi)(X_n)|^2 = u^{(n)}_{\Lambda_L}(\psi)$$

and the desired result is obtained.

Combining this estimate and 2.1.13. we have

$$h^{(n)}_{\infty,\Lambda_{L+2b}}(V\psi) \leq h^{(n)}_{0,\Lambda_L}(\psi) + \frac{2n\nu}{b^2}\|\psi\|^2 .$$

The proof of the proposition now follows that of 2.1.14.

Proposition 2.2.11. The estimates of 2.1.15. and 2.1.16. remain valid.

The estimate of 2.1.15. is obvious because

$$h^{(n)}_{0,\Lambda} \geq h^{(n)}_{\sigma,\Lambda} \qquad , \sigma < 0 .$$

The estimate of 2.1.16. follows by an identical argument but supplemented by the condition

$$U(X_n \cup Y_m) \geq U(X_n) + U(Y_m) .$$

Proposition 2.2.12. The following limit exists for $\mu < -\nu\sigma^2$ and $\beta > 0$

$$P^\sigma(\beta,\mu) = \lim_{L^{(1)},\ldots,L^{(\nu)}\to\infty} \frac{1}{V(\Lambda)} \log Z^\sigma_\Lambda(\beta,\mu) .$$

Moreover

$$P^\sigma(\beta,\mu) = P^0(\beta,\mu) = P^\infty(\beta,\mu) .$$

The proof is identical to that of 2.1.17. but is now based on 2.2.11. and 2.2.9.

2.2.13. We have established that the thermodynamic pressure is independent of the elastic boundary conditions used in its definition. Generalization to varying elasticities can be dealt with in the manner described in 2.1.18.

Further the same pressure is obtained if a kinetic energy operator corresponding to periodic boundary conditions is used to define the total Hamiltonian again as a consequence of 2.1.18. In the next section we consider the situation where the interaction is periodized.

§3 - POTENTIAL AND FINITE RANGE INTERACTIONS

2.3.1. In the sequel we will need stronger conditions on the interactions than we have used up to present. Note that if the interaction U does not take the value $+\infty$ we may write

$$U(X) = \sum_{Y\subset X} \phi(Y)$$

and the function ϕ is uniquely determined, through recursion, by U. The function ϕ is usually referred to as a potential. The properties of ϕ and U are related as follows. If U is invariant, i.e. if

$$U(X+a) = U(X) \qquad , a \in R^\nu , X \subset R^\nu$$

then ϕ is invariant and vice-versa. If $\phi \geq 0$ i.e. if

$$\phi(Y) \geq 0 \qquad , Y \subset R^\nu$$

then $U \geq 0$ but the converse is not true.

We will now consider interactions U_ϕ which are determined by potentials ϕ such that

1. $\phi \geq 0$
2. ϕ is invariant
3. ϕ is decreasing in the sense of 2.2.3.
4. The interaction U_ϕ given by

$$U_\phi(X) = \sum_{Y \subset X} \phi(Y)$$

has the properties demanded of U in 2.2.2. Note that condition 1 automatically implies that

$$U_\phi(X \cup Y) \geq U_\phi(X) + U_\phi(Y) .$$

We will call interactions with the above properties positive potential interactions. They form a subclass of the interactions considered in section §2 and all the results of this section are valid for such interactions.

2.3.2. Next we consider a special subclass of the potential interactions namely those determined by a potential of finite range. The potential ϕ is defined to have range r_ϕ if

$$\phi(X) = 0$$

whenever the diameter of the set X is greater than r_ϕ .

Note that if ϕ is of finite range then

$$U_\phi(X \cup Y) = U_\phi(X) + U_\phi(Y)$$

whenever the distance between the sets X and Y is greater than r_ϕ . In fact for most of our later purposes we could take this latter property as the definition of a finite range interaction and avoid introducing the potentials.

In the case of finite range interactions a number of extra results can easily be deduced form the material of §1 and §2 .

2.3.4. If U_ϕ is a positive potential interaction then it can be combined, in the manner of §2 , with any of the kinetic energy operators of §1 to provide a total Hamiltonian. This Hamiltonian can then in turn be used to define partition functions $Q^\sigma_\Lambda(\beta, n, \phi)$, $Z^\sigma_\Lambda(\beta, z, \phi)$ etc. in the manner of 2.1.1. The first result we can deduce from the estimates of section §2 concerns the free energy

$$F^\sigma_\Lambda(\beta, n, \phi) = \frac{1}{V(\Lambda)} \log Q^\sigma_\Lambda(\beta, n, \phi)$$

in the case that ϕ is of finite range.

Proposition 2.3.5. Let U_ϕ be a positive potential interaction, as defined in 2.3.1., with ϕ of finite range. If $\sigma \geq 0$ then the following limits exist and are equal

$$F(\beta, \rho, \phi) = \lim_{\substack{L^{(1)}, \dots, L^{(\nu)} \to \infty \\ n/V(\Lambda) \to \rho}} \frac{1}{V(\Lambda)} \log Q^\sigma_\Lambda(\beta, n, \phi)$$

$$= \lim_{\substack{L^{(1)}, \dots, L^{(\nu)} \to \infty \\ n/V(\Lambda) \to \rho}} \frac{1}{V(\Lambda)} \log Q^\infty_\Lambda(\beta, n, \phi) .$$

The free energy $F(\beta, \rho, \phi)$ is a convex function of β and a concave function of ρ ; it is connected to the pressure $P(\beta, z, \phi)$ defined in section §2 by the thermodynamic relation

$$P(\beta, z, \phi) = \sup_{\rho \geq 0} \left[\rho \log z + F(\beta, \rho, \phi) \right] .$$

The proof of the existence and equality of the limits uses a different tactic to that of section §2 and starts from the Fisher-Ruelle inequality (cf. 2.1.5.) Let Λ_1 and Λ_2 be disjoint subsets of Λ then

$$Q^\infty_\Lambda(\beta, n, \phi) \geq Q^\infty_{\Lambda_1 \cup \Lambda_2}(\beta, n, \phi) \geq \sum_{m=0}^{n} Q^\infty_{\Lambda_1}(\beta, m, \phi) Q^\infty_{\Lambda_2}(\beta, n-m, \phi) \quad (*)$$

$$\geq Q^\infty_{\Lambda_1}(\beta, m, \phi) Q^\infty_{\Lambda_2}(\beta, n-m, \phi) , 0 \leq m \leq n$$

which is true whenever the distance between Λ_1 and Λ_2 is greater than r_ϕ This inequality is proved in the same manner as in 2.1.5. and together with the positivity and invariance of Q^∞_Λ is sufficient to deduce that the second limit of the proposition exists.

Secondly one uses the estimates of 2.1.14. and 2.2.10. to deduce that

$$Q^{\infty}_{\Lambda_L}(\beta, n, \phi) \leq Q^{0}_{\Lambda_L}(\beta, n, \phi) \leq Q^{\infty}_{\Lambda_{L+2b}}(\beta, n, \phi)\, e^{2n\nu/b^2} .$$

It then follows, as in 2.1.14., that the first limit of the proposition exists for $\sigma = 0$ and is equal to the second. But for $\sigma \geqslant 0$ one has

$$Q^{\infty}_{\Lambda}(\beta, n, \phi) \leq Q^{\sigma}_{\Lambda}(\beta, n, \phi) \leq Q^{0}_{\Lambda}(\beta, n, \phi)$$

from the ordering of the Hamiltonians, and hence the result follows for all $\sigma \geqslant 0$.

The convexity of F as a function of β follows straightforwardly from 1.2.15. and the concavity in ρ can be easily deduced from $(*)$ and the existence of the limit.

We will not prove the thermodynamic relation but refer the reader to the proof given in the book of Ruelle (see bibliography).

2.3.6. Next let us discuss the free energy and the pressure defined with periodic boundary conditions and a periodized potential. The concept of a periodized potential is introduced as follows. Let Λ be a parallelepiped of sides $L^{(1)}, \dots, L^{(\nu)}$ and define n_i^{Λ} by

$$n_i^{\Lambda} = (n_i^{(1)} L^{(1)}, \dots, n_i^{(\nu)} L^{(\nu)})$$

where $n_i^{(1)}, \dots, n_i^{(\nu)}$ are integers. Then if $X_m = \{x_1, \dots, x_m\}$ and ϕ is a positive finite range potential we define the periodized potential by

$$\tilde{\phi}(X_m) = \sum_{n_2^{\Lambda}, \dots, n_m^{\Lambda}} \phi(x_1, x_2 + n_2^{\Lambda}, \dots, x_m + n_m^{\Lambda})$$

where each summation extends over the different possible choices of $\{n_i^{(1)}, \dots, n_i^{(\nu)}\}$ (Note that as ϕ is of finite range the above sum is finite). Let $U_{\tilde{\phi}}$ be the interaction defined now in terms of the periodized potential and then from the positivity of ϕ we note that

$$U_{\phi}(X) \leq U_{\tilde{\phi}}(X) \qquad X \subset \Lambda .$$

If $U^{(n)}_{\tilde{\phi},\Lambda}$ is the operator on $\mathcal{H}^{(n)}(\Lambda)$ such that

$$(U^{(n)}_{\tilde{\phi},\Lambda} \psi)(X_n) = U_{\tilde{\phi}}(X_n)\, \psi(X_n)$$

then we will assume that the generalized sum

$$H^{(n)}_{\Lambda}(\theta) = T^{(n)}_{\Lambda}(\theta) \dotplus U^{(n)}_{\tilde{\phi},\Lambda}$$

is defined. $T^{(n)}_{\Lambda}(\theta)$ is the kinetic energy operator corresponding to the periodic boundary conditions, e.g.

$$\psi(x_1,\ldots,x_n)\Big|_{x_i^{(j)}=0} = e^{i\theta_j}\,\psi(x_1,\ldots,x_n)\Big|_{x_i^{(j)}=L^{(j)}}$$

for all $i = 1,\ldots,n$, $j = 1,\ldots,\nu$ where $0 \leq \theta_i < 2\pi$

We denote by $\tilde{Q}^{\theta}_{\Lambda}(\beta,n,\phi)$ and $\tilde{Z}^{\theta}_{\Lambda}(\beta,z,\phi)$ the partition functions defined with $H^{(n)}_{\Lambda}(\theta)$.

<u>Proposition 2.3.7.</u> <u>Let U_ϕ be a positive potential interaction with finite range and assume that the periodized partition functions above are defined. The following limits exist</u>

$$F(\beta,\rho,\phi) = \lim_{\substack{L^{(1)},\ldots,L^{(\nu)}\to\infty \\ n/V(\Lambda)\to\rho}} \frac{1}{V(\Lambda)} \log \tilde{Q}^{\theta}_{\Lambda}(\beta,n,\phi)$$

$$P(\beta,z,\phi) = \lim_{L^{(1)},\ldots,L^{(\nu)}\to\infty} \frac{1}{V(\Lambda)} \log \tilde{Z}^{\theta}_{\Lambda}(\beta,z,\phi)$$

<u>and are respectively equal to the free energy, defined in 2.3.5., and the pressure, defined in2.2.12., with elastic or repulsive boundary conditions.</u>

The proof follows by, majorizing the partition functions with periodic boundary conditions, minimizing by functions with repulsive boundary conditions, and using 2.3.5.

We have from the discussion of 1.2.17., generalized to n-particles, that

$$T^{(n)}_{0,\Lambda} \leq T^{(n)}_{\Lambda}(\theta)$$

and from 2.3.6. that

$$U^{(n)}_{\phi,\Lambda} \leq U^{(n)}_{\tilde{\phi},\Lambda} .$$

Thus taking the generalized sum one finds

$$H^{(n)}_{0,\Lambda} \leq H^{(n)}_{\Lambda}(\theta)$$

where the Hamiltonian on the left is defined with elastic boundary conditions and potential ϕ whilst the Hamiltonian on the right uses periodic boundary conditions and $\tilde{\phi}$. It now follows directly from 1.2.15. that

$$Q^{0}_{\Lambda}(\beta,n,\phi) \geq \tilde{Q}^{\theta}_{\Lambda}(\beta,n,\phi) .$$

Secondly assume that the parallelepiped is large enough that $L^{(i)} > r_\phi$. If $X \subset \Lambda$ is such that the distance between X and the surface of Λ is greater than $r_\phi/2$ then we have

$$U_\phi(x) = U_{\tilde\phi}(x)$$

as a consequence of the definition of $\tilde\phi$ and the range of ϕ. Now if $\psi \in D(h_\Lambda^{(n)}(\theta,\cdot))$ and $\psi(x) = 0$ whenever the distance between x and the surface of Λ is less than $r_\phi/2$ it follows that $\psi \in D(h_{\infty,\Lambda}^{(n)})$ and

$$h_{\infty,\Lambda}^{(n)}(\psi) = h_\Lambda^{(n)}(\theta,\psi) .$$

($h_\Lambda^{(n)}(\theta,\cdot)$ denotes the form associated with $H_\Lambda^{(n)}(\theta)$). However if Λ' denotes the parallelepiped formed from Λ by omitting the set of points which are at a distance less than $r_\phi/2$ from the surface of Λ we have for the above ψ that

$$h_{\infty,\Lambda}^{(n)}(\psi) = h_{\infty,\Lambda'}^{(n)}(\psi) .$$

But such ψ form a core of $h_{\infty,\Lambda'}^{(n)}$ and hence

$$\tilde Q_\Lambda^\theta(\beta,n,\phi) \geq Q_{\Lambda'}^\infty(\beta,n,\phi)$$

by part o1 of 1.2.15. The first statement of the proposition then follows from these estimates and 2.3.5. ; the second statement is proved similarly.

GENERAL BIBLIOGRAPHY

A description of the Fisher-Ruelle discussion of the thermodynamic pressure with repulsive boundary conditions is given in

D. RUELLE

Statistical Mechanics - Benjamin (New York) 1969.

For the case of elastic boundary conditions we have followed the methods applied to hard core particles in

D.W. ROBINSON

Commun. Math. Phys. 16 , 290 (1970)

In this latter setting the estimate analogous to 2.1.13. is to be found in

R.E. GRIFFITHS and D.W. ROBINSON

(to be published).

The discussion of bose hard core particles with purely elastic boundary conditions at low density is carried through by the methods of functional integration in

I.D. NOVIKOV

Funct. Anal. and Appl. 3 71 (1969)

See also

J. GINIBRE

Lectures delivered at les Houches (1970)

EXERCISES

1. Let $f(a^{(1)},..,a^{(\nu)})$ be defined and real for $a^{(1)},..,a^{(\nu)} \in R$ $a^{(1)}>0,..,a^{(\nu)}>0$ and subadditive separately in each variable, i.e.

$$f(a^{(1)},..,b^{(i)}+c^{(i)},..,a^{(\nu)}) \leq f(a^{(1)},..,b^{(i)},..,a^{(\nu)}) + f(a^{(1)},..,c^{(i)},..,a^{(\nu)}).$$

Further assume that

$$\sup_{a^{(1)},..,a^{(\nu)}} f(a^{(1)},..,a^{(\nu)}) / \prod_{i=1}^{\nu} a^{(i)} < +\infty$$

prove that

$$\lim_{a^{(1)},..,a^{(\nu)}\to\infty} f(a^{(1)},..,a^{(\nu)})/\prod_{i=1}^{\nu} a^{(i)} = \inf_{a^{(1)},..,a^{(\nu)}} f(a^{(1)},..,a^{(\nu)})/\prod_{i=1}^{\nu} a^{(i)} .$$

[Hint ; for $\epsilon > 0$ choose $b^{(1)},..,b^{(\nu)}$ such that

$$f(b^{(1)},..,b^{(\nu)})/\prod_{i=1}^{\nu} b^{(i)} < \inf_{a^{(1)},..,a^{(\nu)}} f(a^{(1)},..,a^{(\nu)})/\prod_{i=1}^{\nu} a^{(i)} + \epsilon = C+\epsilon .$$

Decompose each $a^{(i)}$ in the form $a^{(i)} = n^{(i)} b^{(i)} + c^{(i)}$ with $n^{(i)}$ integer and $0 \leq c^{(i)} < b^{(i)}$ then note that from the subadditivity and boundedness properties

$$f(a^{(1)},..,a^{(\nu)})/\prod_{i=1}^{\nu} a^{(i)} \leq C+\epsilon + R(a^{(1)},..,a^{(\nu)})$$

where $R \to 0$ as $a^{(1)},..,a^{(\nu)} \to \infty$.]

2. Provide the proofs of propositions 2.1.8., 2.1.9. and 2.3.5.

3. Consider the operator $K^{\mu}_{0,\Lambda}$ of 1.3.4. With sufficient smoothness conditions on Λ_1 and Λ_2 prove

$$K^{\mu}_{0,\Lambda_1\cup\Lambda_2} \geq \left(K^{\mu}_{0,\Lambda_1} \otimes \mathbb{1}_{\Lambda_2} + \mathbb{1}_{\Lambda_1} \otimes K^{\mu}_{0,\Lambda_2}\right)^{**} \quad , \Lambda_1 \cap \Lambda_2 = \phi.$$

Consider the analogous operator $K^{\mu}_{\infty,\Lambda}$ and prove that

$$K^{\mu}_{\infty,\Lambda_1\cup\Lambda_2} \leqslant \left(K^{\mu}_{\infty,\Lambda_1}\otimes \mathbb{1}_{\Lambda_2} + \mathbb{1}_{\Lambda_1}\otimes K^{\mu}_{\infty,\Lambda_2}\right)^{**} \qquad \Lambda_1\cap\Lambda_2=\phi.$$

Hence use the direct definition of the (non-interacting) grand canonical partition functions

$$Z^{0}_{\Lambda}(\beta,\mu) = \mathrm{Tr}_{\mathcal{H}(\Lambda)}\left(e^{-\beta K^{\mu}_{0,\Lambda}}\right) \qquad \text{etc.,}$$

to deduce the sub-multiplicative properties of 2.1.5. and 2.1.6.

CHAPTER III

LOCAL ENTROPY AND ENERGY DENSITY

INTRODUCTION

In the second half of these lectures we wish to consider a different aspect of the thermodynamic pressure namely its definition as the supremum of the entropy at fixed energy density. In this chapter we will consider properties of a fixed finite system Λ and work with the Hilbert space $\mathcal{H} = \mathcal{H}(\Lambda)$. H will represent a 'Hamiltonian' operator on $\mathcal{H}$ with associated form h. The only properties that we will immediately need are that H is positive, self-adjoint, has discrete spectrum with finite multiplicity, and $\exp\{-\beta H\}$ is of trace- class for all $\beta > 0$.

The quantum mechanical states of the system are described by density matrices ρ, on $\mathcal{H}$. A density matrix ρ is a positive self-adjoint operator with domain $\mathcal{H}$ and trace-norm unity i.e.

$$\mathrm{Tr}_{\mathcal{H}}(\rho) = 1 \quad .$$

This last condition implies that the spectrum of ρ is discrete, with a possible accumulation point at zero ; we denote the eigenvalues of ρ, arranged in decreasing order repeated according to multiplicity by $(\lambda_m(\rho))_{m\geq 1}$ and an associated orthonormal family of eigenvectors by $(\psi_m^\rho)_{m\geq 1}$. We denote the set of density matrices by N.

We can associate with each $\rho \in N$ an entropy density by

$$S_\Lambda(\rho) = -\frac{1}{V(\Lambda)} \mathrm{Tr}_{\mathcal{H}}(\rho \log \rho) = \frac{1}{V(\Lambda)} \sum_{m\geq 1} -\lambda_m(\rho) \log \lambda_m(\rho)$$

where we set $-t\log t = 0$ for $t = 0$. Note that $S_\Lambda(\rho) \in [0, +\infty]$. Physically $S_\Lambda(\rho)$ is a measure of the disorder of the state ρ.

Next let N_H be the subclass of N composed of the density matrices such that the ψ_m^ρ can be chosen in $D(h)$ and

$$h_\Lambda(\rho) = \frac{1}{V(\Lambda)} \sum_{m\geq 1} \lambda_m(\rho)\, h(\psi_m^\rho) < +\infty$$

The function $\rho \in N_H \to h_\Lambda(\rho) \in [0,+\infty]$ is the energy density of the state ρ with respect to the Hamiltonian H (with a suitable interpretation of the trace one has

$$h_\Lambda(\rho) = \frac{1}{V(\Lambda)} \mathrm{Tr}_{\mathcal{H}}(\rho H) \qquad).$$

Finally for $\beta > 0$ we introduce the conditional entropy density $\rho \in N_H \to S_\Lambda(\beta,\rho)$ by

$$S_\Lambda(\beta,\rho) = S_\Lambda(\rho) - \beta h_\Lambda(\rho) .$$

As $h_\Lambda(\rho) < +\infty$ this definition is unambigous.

The starting point of our investigation is the following relation

$$P_\Lambda(\beta) = \frac{1}{V(\Lambda)} \log \mathrm{Tr}(e^{-\beta H}) = \sup_{\rho \in N_H} S_\Lambda(\beta,\rho)$$

i.e. the local pressure is the supremum of the conditional entropy. Following the Gibbs interpretation the state (or states) for which the supremum is attained is interpreted as the equilibrium state of the system ; i.e. one adopts the position that equilibrium is described by the state of maximum disorder compatible with a given energy density.

The proof of this relation is simple and consists essentially of two arguments. Firstly one shows that

$$P_\Lambda(\beta) \geq S_\Lambda(\beta,\rho) \qquad , \rho \in N_H$$

and secondly one constructs $\rho^* \in N_H$ such that

$$P_\Lambda(\beta) = S_\Lambda(\beta,\rho^*) .$$

The first step is deduced as follows

$$\begin{aligned} P_\Lambda(\beta) - S_\Lambda(\beta,\rho) &= -\frac{1}{V(\Lambda)} \sum_{m\geq 1} \left[-\lambda_m(\rho)\log\lambda_m(\rho) + \lambda_m(\rho) \log \frac{e^{-\beta h(\psi_m^\rho)}}{\mathrm{Tr}_{\mathcal{H}}(e^{-\beta H})} \right] \\ &\geq \frac{1}{V(\Lambda)} \sum_{m\geq 1} \left[\frac{e^{-\beta h(\psi_m^\rho)}}{\mathrm{Tr}_{\mathcal{H}}(e^{-\beta H})} - \lambda_m(\rho) \right] \\ &\geq 0 \end{aligned}$$

where one successively uses the convexity inequality

$$-x(\log x - \log y) \leq y - x \qquad , x, y > 0$$

and part 1 of 1.2.15. The second step is easy because it suffices to take

$$\rho^* = e^{-\beta H} / \mathrm{Tr}_{\mathcal{H}}(e^{-\beta H}) .$$

In fact with our assumptions it can be proved that ρ^* is the unique state for which the supremum is attained but we will not show this. (cf. Exercise 1).

The derivation of a principle of maximum conditional entropy for the thermodynamic pressure is more complex than the above. As one wishes to consider systems of an arbitrary size it is no longer possible to characterize their states by density matrices on a fixed Hilbert space. One has rather to consider a state given by a family of density matrices $\{\rho_\Lambda ; \Lambda \subset R^\nu\}$, each ρ_Λ characterizing a finite subsystem Λ of the system. Secondly one needs to extend the definition of $S_\Lambda(\beta, \rho)$ to be a function over these states (essentially one has to define the limit of $S_\Lambda(\beta, \rho)$ as $\Lambda \to \infty$ and for this purpose it is technically necessary to consider only those states invariant under space translations). Finally one has to establish that the supremum of the conditional entropy density is exactly the thermodynamic pressure. This last step is in fact the most onerous ; one can use convexity, as above, to show that the pressure is an upper bound but it is more involved to show that the bound is attained. As the states entering the maximum problem are now given in an essentially implicit form there is no direct constructive proof of the type occurring for the finite system. It is necessary to invoke limiting arguments which can only be readily validated if the conditional entropy has an upper semi-continuity property. Thus a large amount of the following material will be devoted to the study of topological properties of the states entering into the maximum problem and semi-continuity properties of the energy density and conditional entropy density.

§1 - TOPOLOGIES OF NORMAL STATES

3.1.1. We consider a complex Hilbert space $\mathcal{H}$ and the family N of density matrices ρ on $\mathcal{H}$. The set of density matrices can be topologized in various manners with the aid of the Hilbert space structure but for our purposes it is convenient to consider the topologies induced by irreducible C^* algebras of bounded operators on $\mathcal{H}$. The basic reason why this is useful is that the C^* algebra can be considered as an abstract object independent of $\mathcal{H}$. This is a natural advantage when we later consider an infinite system because its states are no longer directly connected to one fixed Hilbert space but can all be associated with a suitable C^* algebra.

We first recall a number of basic definitions and results concerning algebras of bounded operators on Hilbert space.

3.1.2. A bounded operator A on $\mathcal{H}$ is an operator with $D(A) = \mathcal{H}$ and

$$\|A\| = \sup_{\psi \in \mathcal{H}} \frac{\|A\psi\|}{\|\psi\|} < +\infty .$$

If A^* is the adjoint of A one has

$$\|A^*\|^2 = \|A\|^2 = \|AA^*\| = \|A^*A\| .$$

A *algebra on $\mathcal{H}$ is a set of bounded operators closed under the operations of addition, multiplication, multiplication by a scalar, and also the adjoint operation.

A C*algebra is a *algebra which is closed with respect to the topology induced by the norm $\|\cdot\|$.

A C* algebra $\mathfrak{A}$ on $\mathcal{H}$ is said to be irreducible if and only if each bounded operator on $\mathcal{H}$ which commutes with all elements of $\mathfrak{A}$ is a multiple of the identity.

The set of all bounded operators on $\mathcal{H}$ forms an irreducible C* algebra on $\mathcal{H}$ which is noted by $\mathcal{L}(\mathcal{H})$. Similarly the set of all compact operators forms an irreducible C* algebra on $\mathcal{H}$ noted by $\mathcal{LC}(\mathcal{H})$.

3.1.3. As each C* algebra forms a complete metric space, a Banach space, one can define its dual. The dual space $\mathfrak{A}^*$ of a C* algebra $\mathfrak{A}$ is the set of all continuous linear functionals ω over $\mathfrak{A}$, i.e. the set of functions ω such that

$$\omega(\alpha_1 A_1 + \alpha_2 A_2) = \alpha_1 \omega(A_1) + \alpha_2 \omega(A_2)$$

for all $\alpha_1, \alpha_2 \in \mathbb{C}$, $A_1, A_2 \in \mathfrak{A}$, and with

$$\|\omega\| = \sup_{A \in \mathfrak{A}} |\omega(A)/\|A\|| < +\infty .$$

Note that $\mathfrak{A}^*$ is complete with respect to the metric topology defined by the norm $\omega \to \|\omega\|$, i.e. $\mathfrak{A}^*$ is a Banach space.

The metric topology induced on $\mathfrak{A}^*$ by the norm $\|\cdot\|$ is referred to as the <u>uniform topology</u>. A complete set of neighbourhoods of $\omega \in \mathfrak{A}^*$ for this topology is given by

$$\mathcal{U}(\omega; \epsilon) = \{\omega' ; \omega' \in \mathfrak{A}^*, \|\omega' - \omega\| < \epsilon\}$$

where $\epsilon > 0$.

The dual space $\mathfrak{A}^*$ can also be equipped in a standard way with a <u>weak* topology</u>. A complete set of neighbourhoods of $\omega \in \mathfrak{A}^*$ for this topology is given by

$$\mathcal{W}(\omega; A_1, .., A_n, \epsilon) = \{\omega' ; \omega' \in \mathfrak{A}^*, |\omega'(A_i) - \omega(A_i)| < \epsilon, i = 1, .., n\}$$

where $A_1, .., A_n \in \mathfrak{A}$ and $\epsilon > 0$.

The uniform topology is finer than the weak* topology ; each neighbouhood of the set $\mathcal{W}$ contains a neighbourhood of the set $\mathcal{U}$.

<u>3.1.4.</u> In order to analyse properties of subsets of $\mathfrak{A}^*$, or any general topological space, it is useful to introduce the notion of a <u>directed set</u>, or generalized sequence.

A set I is <u>partially ordered</u> when for certain pairs $\alpha, \beta \in I$ an order relation is given $\alpha \leq \beta$ which is <u>reflexive</u> ($\alpha \leq \alpha$), <u>transitive</u> ($\alpha \leq \beta$ and $\beta \leq \gamma$ imply $\alpha \leq \gamma$) and <u>antisymmetric</u> ($\alpha \leq \beta$ and $\beta \leq \alpha$ imply $\alpha = \beta$). A partially ordered set I is called <u>directed</u> when for each pair $\alpha, \beta \in I$ there is always a $\gamma \in I$ such that $\alpha \leq \gamma$ and $\beta \leq \gamma$. Let I be a directed set and M an arbitrary subset of a topological space whose elements ω_α are indexed by the $\alpha \in I$; then the ω_α form a directed set in M. The directed set ω_α, $\alpha \in I$ is said to be convergent to $\omega \in M$ if for each neighbourhood U of ω there is an $\alpha(U) \in I$ such that $\omega_\alpha \in U$ for all

$\alpha \geq \alpha(U)$; ω is the limit of ω_α, $\omega_\alpha \to \omega$.

A subset M of a topological space is closed if and only if it contains the limits of all directed sets constructed from its elements ; M is compact if and only if each directed set in M has at least one limit point in M .

Each compact set is closed and each closed subset of a compact set is compact.

A directed set $\omega_\alpha \in M$ is called a Cauchy system if for each neighbourhood $U \in M$ there is an $\alpha(U) \in I$ such that $\omega_\alpha, \omega_\beta \in U$ for all $\alpha, \beta \geq \alpha(U)$.

A subset M of a topological space is complete if and only if each Cauchy system constructed from its elements has a limit point in M .

Each closed subset of a complete set is complete.

3.1.5. We will also need to consider continuity properties of functions over topological spaces. Let M be such a space and $\omega \in M \to f(\omega) \in [-\infty, +\infty]$ a real function over M .

The function f is defined to be lower semi-continuous if for each $x \in [-\infty, +\infty)$ the set $\{\omega ; \omega \in M, f(\omega) \leq x\}$ is closed (or, equivalently, if the set $\{\omega ; \omega \in M , f(\omega) > x\}$ is open).

Two other criteria for lower semi-continuity are the following ; f is lower semi-continuous if and only if for each ω such that $f(\omega) > x$ there is a neighbourhood $U(\omega)$ with the property $f(\omega') > x$ for all $\omega' \in U(\omega)$

f is lower semi-continuous if and only if

$$f(\omega) \leq \liminf_{\omega_\alpha \to \omega} f(\omega_\alpha)$$

or, introducing the explicit definition of the inferior limit, if and only if

$$f(\omega) \leq \sup_{\alpha} \inf_{\beta \geq \alpha} f(\omega_\beta) .$$

The sum of a finite family of lower semi-continuous functions is lower semi-continuous. The upper envelope of a family of lower semi-continuous functions is lower semi-continuous.

A function f is said to be upper semi-continuous if $-f$ is lower semi-continuous. A function f is continuous if and only if it is upper and lower semi-continuous.

The following two points will be of use in considering the maximum principle for the conditional entropy density.

An upper semi-continuous function defined over a compact set attains its supremum.

An affine upper semi-continuous function defined over a convex compact set attains its supremum at an extremal point.

3.1.6. Let us now return to the discussion of the dual space $\mathfrak{A}^*$ of the C* algebra $\mathfrak{A}$.

An element $\omega \in \mathfrak{A}^*$ which is positive

$$\omega(A^*A) \geq 0 \qquad , \quad A \in \mathfrak{A}$$

and normalized

$$\|\omega\| = 1$$

is called a state. The set of states over $\mathfrak{A}$ will be denoted by $E_{\mathfrak{A}}$.

$E_{\mathfrak{A}}$ is a convex subset of $\mathfrak{A}^*$ which is in general relatively compact in the weak* topology, i.e. $E_{\mathfrak{A}}$ has a compact closure. If $\mathfrak{A}$ has an identity element then $E_{\mathfrak{A}}$ is weak* compact. $E_{\mathfrak{A}}$ is closed and hence complete in the uniform topology.

3.1.7. Let $\mathfrak{A}$ be an irreducible C* algebra on $\mathcal{H}$. The set N of density matrices on $\mathcal{H}$ determines a subset, also denoted by N, of $E_{\mathfrak{A}}$ through the correspondence $\rho \in N \to \omega_\rho \in E_{\mathfrak{A}}$ where

$$\omega_\rho(A) = \mathrm{Tr}_{\mathcal{H}}(\rho A) \qquad , \quad A \in \mathfrak{A} .$$

The set of states N determined in this fashion are called normal states. If $\mathfrak{A} = \mathcal{LC}(\mathcal{H})$ then $N = E_{\mathfrak{A}}$ but in general N is a proper subset of $E_{\mathfrak{A}}$ which is relatively compact in the weak* topology and closed in the uniform topology.

3.1.8. We will have occasion to use the following form of the Kaplansky density theorem.

Let $\mathfrak{A}$ be an irreducible C* algebra, E a finite dimensional projection operator, and B a bounded operator, on the Hilbert space $\mathcal{H}$. For $\epsilon > 0$ there is an $A \in \mathfrak{A}$ such that

$$\|(A-B)E\| < \epsilon$$

and

$$\|A\| = \|B\| .$$

3.1.9. Let us now examine the uniform topology on N with $\mathfrak{A}$ irreducible. For $\omega_\rho, \omega_\sigma \in N$ one has

$$\|\omega_\rho - \omega_\sigma\| = \sup_{A \in \mathfrak{A}, \|A\|=1} |\omega_\rho(A) - \omega_\sigma(A)|$$

$$= \sup_{A \in \mathfrak{A}, \|A\|=1} |Tr_{\mathcal{H}}((\rho-\sigma)A)| .$$

Now recall that the trace-norm of a compact operator C on $\mathcal{H}$ is defined by

$$|||C||| = \sup_{A \in \mathcal{L}(\mathcal{H}), \|A\|=1} |Tr_{\mathcal{H}}(CA)| .$$

But as the spectrum of C is discrete with a possible accumulation point at zero one can deduce from 3.1.8. that the supremum in the last expression can be taken over any irreducible sub C^*-algebra of $\mathcal{L}(\mathcal{H})$. Thus

$$\|\omega_\rho - \omega_\sigma\| = |||\rho - \sigma|||$$

i.e. the trace-norm topology on the density matrices coincides with the uniform topology on N . Alternatively stated the uniform topology on N is independent of the particular C^*-algebra $\mathfrak{A}$ used in its definition.

3.1.10. A useful result concerning the convergence of compact operators is the following : let ρ and σ be two density matrices and $(\lambda_m(\rho))_{m\geq 1}, (\lambda_m(\sigma))_{m\geq 1}$ their eigenvalues arranged in decreasing order repeated according to multiplicity then

$$|\lambda_m(\rho) - \lambda_m(\sigma)| \leq \|\rho - \sigma\| \leq |||\rho - \sigma||| = \|\omega_\rho - \omega_\sigma\| \quad , m = 1, 2, \ldots .$$

Further if $\Delta = (\delta_1, \delta_2)$ is an interval such that the extremities do not belong to the spectrum of ρ and $E_\rho(\Delta)$, $E_\sigma(\Delta)$ are the corresponding spectral projectors of ρ and σ respectively then there is a $C(\Delta) > 0$ such that

$$\|E_\rho(\Delta) - E_\sigma(\Delta)\| < C(\Delta)\|\rho - \sigma\| \leq C(\Delta)\|\omega_\rho - \omega_\sigma\|$$

for $\|\rho - \sigma\|$ small.

§2 - LOCAL ENERGY AND ENTROPY DENSITIES

3.2.1. We will now return to the discussion of the energy density and conditional entropy density defined in the introduction. Throughout the rest of this chapter we consider $\Lambda \subset R^{\nu}$ fixed and omit it as an index ; similarly we omit the factor $V(\Lambda)$. We extend the definitions of the introduction by choosing an irreducible C* algebra $\mathfrak{A}$ on $\mathcal{H}$ and defining the energy and entropy densities as functions over the complete state space $E_{\mathfrak{A}}$ of $\mathfrak{A}$. We first consider the energy density.

3.2.2. Let H be a lower semi-bounded self-adjoint operator on $\mathcal{H}$ with discrete spectrum with finite multiplicites for each of the eigenvalues. Denote the spectral representation of H by

$$H = \int dE_H(\lambda)\, \lambda$$

and note that for each $m \in (-\infty, +\infty)$ the operator

$$H_m = \int_{\lambda \leq m} dE_H(\lambda)\, \lambda$$

is in $\mathcal{L}(\mathcal{H})$. The energy density $\hat{H}$ is defined as a functional over the states $E_{\mathfrak{A}}$ of $\mathfrak{A}$ by

$$\hat{H}(\omega) = +\infty \qquad , \quad \omega \notin N$$

$$\hat{H}(\omega_\rho) = \sup_m \omega_\rho(H_m) \qquad , \quad \omega \in N \;.$$

Note that the definition of $\omega_\rho \in N \rightarrow \hat{H}(\omega_\rho)$ coincides with the definition of the function given in the introduction.

This can be seen as follows. First, by addition of a constant, H can be adjusted to be positive and then both definitions are equivalent to partial resummations of

$$\sum_{m,n} \lambda_m(H)\, \lambda_n(\rho)\, |(\psi_n^\rho, \psi_m)|^2 \,.$$

But the order of summation is irrelevant because all terms are positive and hence the two definitions coincide. We have used the notation $(\lambda_m(H))_{m \geq 1}$ for the eigenvalues of H arranged in increasing order repeated according to multiplicity and $(\psi_m)_{m \geq 1}$ for a corresponding set of eigenvectors. Note that one also has

$$\hat{H}(\omega_\rho) = \sum_{m\geq 1} \lambda_m(H)(\psi_m, \rho\psi_m) . \qquad (*)$$

Proposition 3.2.3. The function $\omega \in E_{\mathfrak{A}} \rightarrow \hat{H}(\omega) \in R$ is affine and lower semi-continuous in the weak* topology.

The subsets of the normal states defined by

$$D_n(H) = \{\omega \,;\, \omega \in E_{\mathfrak{A}} \,,\, \hat{H}(\omega) \leq n\} \qquad n \in R$$

are convex, closed in the weak* topology, and the weak* topology and uniform topologies coincide on $D_n(H)$.

The function $\hat{H}$ is affine, i.e.

$$\hat{H}(\alpha\omega_1 + (1-\alpha)\omega_2) = \alpha\,\hat{H}(\omega_1) + (1-\alpha)\hat{H}(\omega_2)$$

for $0 < \alpha < 1$ and $\omega_1, \omega_2 \in E_{\mathfrak{A}}$, as a direct consequence of (*) and the fact that $\hat{H}(\omega_1) = +\infty$ implies that $\hat{H}(\alpha\omega_1 + (1-\alpha)\omega_2) = +\infty$. The convexity of the $D_n(H)$ then follows immediately. We will deduce that $\hat{H}$ is lower semi-continuous from the fact that the $D_n(H)$ are weak* closed (cf. 3.1.5.) Thus the rest of the proof is based upon the analysis of the sets $D_n(H)$.

First note that if $\omega_\rho, \omega_\sigma \in N$ then

$$|\omega_\rho(H_m) - \omega_\sigma(H_m)| = |Tr_{\mathcal{H}}((\rho-\sigma)H_m)|$$
$$\leq \|H_m\| \, \|\omega_\rho - \omega_\sigma\|$$

Thus the function $\omega \in N \rightarrow \hat{H}(\omega) \in R$ is defined as the upper envelope of a family of functions which are continuous in the uniform topology on N and as a consequence is lower semi-continuous in this topology (3.1.5.). It then follows that the $D_n(H)$ are closed in the uniform topology on N. But N is a uniformly closed subset of $E_{\mathfrak{A}}$ (3.1.7.) and hence the $D_n(H)$ are closed in the uniform topology on $E_{\mathfrak{A}}$ (cf. definition of closed 3.1.4.). Note that the $D_n(H)$ are automatically complete in the uniform topology as a result of being closed (3.1.4.).

The rest of the proof is now based upon the following estimate. Let E_λ be the finite dimensional projector on the subspace of $\mathcal{H}$ spanned by the eigenvectors of H which correspond to eigenvalues which are smaller than $\lambda \in R$.

For $\omega_\rho \in D_n(H)$ we then have

$$n \geq \hat{H}(\omega_\rho) \geq \sum_{\lambda_m(H) \geq \lambda} \lambda_m(H)(\psi_m, \rho\,\psi_m)$$

$$\geq \lambda \sum_{\lambda_m(H) \geq \lambda} (\psi_m, \rho\,\psi_m)$$

$$= \lambda \; \mathrm{Tr}_{\mathcal{H}}(\rho(1-E_\lambda))$$

$$= \lambda \; \omega_\rho(1-E_\lambda) \; .$$

Thus

$$0 \leq \omega_\rho(1-E_\lambda) \leq n/\lambda$$

uniformly for all $\omega_\rho \in D_n(H)$ (cf. the similar estimate used in 1.2.10).

Now to prove that the uniform topology and weak* topology coincide on $D_n(H)$ we must show that each neighbourhood of the basis defining the uniform topology contains a neighbourhood of the basis defining the weak* topology. In this case the two topologies are simultaneously finer and coarser, i.e. they are equal.

Now given $\epsilon > 0$ choose λ such that $4(n/\lambda)^{1/2} < \epsilon/5$. Next fix $\omega_\rho \in D_n(H)$. As E_λ is a finite dimensional projector we can certainly choose a finite number p, dependent on λ, of operators $B_1, \ldots, B_p \in \mathcal{L}(\mathcal{H})$ with $\|B_1\| = \ldots = \|B_p\| = 1$ and such that if

$$|\omega_\sigma(E_\lambda B_i E_\lambda) - \omega_\rho(E_\lambda B_i E_\lambda)| < 4\epsilon/5 \quad , \quad i = 1, \ldots, p$$

then

$$\sup_{B \in \mathcal{L}(\mathcal{H})} |\omega_\sigma(E_\lambda B E_\lambda) - \omega_\rho(E_\lambda B E_\lambda)| < 4\epsilon/5 \; .$$

But using 3.1.8. and the irreducibility of $\mathfrak{A}$ we can then choose $A_1, \ldots, A_p \in \mathfrak{A}$ with $\|A_1\| = \ldots = \|A_p\| = 1$ such that

$$\|(A_i - B_i)E_\lambda\| < \epsilon/5 \qquad i = 1, \ldots, p \; .$$

However for an arbitrary $B \in \mathcal{L}(\mathcal{H})$ and all pairs $\omega_\sigma, \omega_\tau \in D_n(H)$ we have

$$\big|\,|\omega_\sigma(B) - \omega_\tau(B)| - |\omega_\sigma(E_\lambda B E_\lambda) - \omega_\tau(E_\lambda B E_\lambda)|\,\big|$$

$$\leq |\omega_\sigma(B) - \omega_\sigma(E_\lambda B E_\lambda) - \omega_\tau(B) + \omega_\tau(E_\lambda B E_\lambda)|$$

$$\leq |\omega_\sigma(B) - \omega_\sigma(E_\lambda B E_\lambda)| + |\omega_\tau(B) - \omega_\tau(E_\lambda B E_\lambda)|$$

$$= |\omega_\sigma((1-E_\lambda)B) + \omega_\sigma(E_\lambda B(1-E_\lambda))|$$

$$+ |\omega_\tau((1-E_\lambda)B) + \omega_\tau(E_\lambda B(1-E_\lambda))|$$

$$\leq 4(n/\lambda)^{\frac{1}{2}} \|B\| \quad .$$

Finally consider the weak* neighbourhood of ω_ρ given by

$$\mathcal{W}(\omega_\rho\,;\, A_1,\ldots, A_p\,,\, \epsilon/5) = \{\omega_\sigma\,;\, \omega_\sigma \in D_n(H)\,,\, |\omega_\sigma(A_i) - \omega_\rho(A_i)| < \epsilon/5\} \,.$$

If $\omega_\sigma \in \mathcal{W}$ we then have

$$|\omega_\sigma(E_\lambda B_i E_\lambda) - \omega_\rho(E_\lambda B_i E_\lambda)| < 2\epsilon/5 + |\omega_\sigma(E_\lambda A_i E_\lambda) - \omega_\rho(E_\lambda A_i E_\lambda)|$$

$$< 3\epsilon/5 + |\omega_\sigma(A_i) - \omega_\rho(A_i)|$$

$$< 4\epsilon/5 \quad .$$

Thus

$$\|\omega_\sigma - \omega_\rho\| = \sup_{B \in \mathcal{L}(\mathcal{H})} |\omega_\sigma(B) - \omega_\rho(B)| / \|B\|$$

$$< \epsilon/5 + \sup_{B \in \mathcal{L}(\mathcal{H})} |\omega_\sigma(E_\lambda B E_\lambda) - \omega_\rho(E_\lambda B E_\lambda)| / \|B\|$$

$$< \epsilon$$

i.e. we have concluded that the weak* neighbourhood $\mathcal{W}(\omega_\rho\,;\, A_1,\ldots, A_p\,,\, \epsilon/5)$ is contained in the uniform neighbourhood

$$\mathcal{U}(\omega_\rho\,;\,\epsilon) = \{\omega_\sigma\,;\, \omega_\sigma \in D_n(H)\,,\, \|\omega_\sigma - \omega_\rho\| < \epsilon\} \,.$$

Thus we have proved that the weak* topology and the uniform topology induced on $D_n(H)$ coincide. It remains to prove that $D_n(H)$ is closed in the weak* topology.

Let ω_{ρ_α} be a directed set of states in $D_n(H)$ which converges in the weak* topology to a state $\omega \in E_{\mathfrak{A}}$. Thus (ω_{ρ_α}) is a Cauchy system for the weak* topology, e.g. given $\epsilon > 0$, $A_1, \ldots, A_p \in \mathfrak{A}$ with $\|A_1\| = \ldots = \|A_p\| = 1$ there is an α such that

$$|\omega_{\rho_\beta}(A_i) - \omega_{\rho_\gamma}(A_i)| < \epsilon/5 \quad , i = 1,\ldots,p \quad \text{and} \quad \beta,\gamma > \alpha \ .$$

But then the above estimates give

$$\| \omega_{\rho_\beta} - \omega_{\rho_\gamma} \| < \epsilon \qquad , \quad \beta,\gamma > \alpha$$

i.e. (ω_{ρ_α}) is a cauchy system for the uniform topology. However we have already proved that $D_n(H)$ is complete in the uniform topology. Thus (ω_{ρ_α}) must have a uniform limit point $\omega' \in D_n(H)$. But as the uniform topology on $E_{\mathfrak{A}}$ is finer than the weak* topology we must have $\omega' = \omega$ i.e. the weak* limit point ω is in $D_n(H)$. This proves that $D_n(H)$ is closed in the weak* topology.

3.2.4. Next we consider the conditional entropy. We will now assume that the operator H has the property

$$\mathrm{Tr}_{\mathcal{H}}\left(e^{-\beta H}\right) < +\infty$$

for all $\beta > 0$, a condition which was not necessary for the considerations of the last paragraphs.

The convexity argument given in the introduction shows that if $\omega_\rho \in D(H)$ where $D(H) \subset N$ is defined by

$$D(H) = \{ \omega \ ; \ \omega \in E_{\mathfrak{A}} \ , \ \hat{H}(\omega) < +\infty \}$$

then

$$-\mathrm{Tr}(\rho \log \rho) \leq \beta \hat{H}(\omega_\rho) + \log \mathrm{Tr}_{\mathcal{H}}\left(e^{-\beta H}\right) \qquad , \quad \beta > 0$$
$$< +\infty \ .$$

Thus we can unambigously define the conditional entropy as a function over the state space $E_{\mathfrak{A}}$ of $\mathfrak{A}$ by

$$S(\beta H ; \omega_\rho) = -\mathrm{Tr}_{\mathcal{H}}(\rho \log \rho) - \beta \hat{H}(\omega_\rho) \qquad , \quad \omega_\rho \in D(H)$$
$$S(\beta H ; \omega) = -\infty \qquad , \quad \omega \in E_{\mathfrak{A}} \ , \ \omega \notin D(H)$$

for all $\beta > 0$. With this definition we then have

$$S(\beta H ; \cdot) \in \left[-\infty \ , \ \log \mathrm{Tr}_{\mathcal{H}}\left(e^{-\beta H}\right) \right] .$$

<u>Proposition 3.2.5.</u> <u>For each $\beta > 0$ the function $\omega \in E_{\mathfrak{A}} \to S(\beta H ; \omega)$ is upper semi-continuous in the weak* topology and satisfies the inequalities</u>

$$S(\beta H ; \alpha\omega_1 + (1-\alpha)\omega_2) \geq \alpha S(\beta H ; \omega_1) + (1-\alpha) S(\beta H ; \omega_2)$$

$$S(\beta H ; \alpha\omega_1 + (1-\alpha)\omega_2) \leq \alpha S(\beta H ; \omega_1) + (1-\alpha) S(\beta H ; \omega_2) + \log 2$$

where $0 < \alpha < 1$ and $\omega_1, \omega_2 \in E_{\mathfrak{A}}$.

The subsets of the normal states defined by

$$D_n(S; \beta H) = \{ \omega \; ; \; \omega \in E_{\mathfrak{A}} \, , \; S(\beta H; \omega) \geq n \} \qquad n \in R$$

are convex, closed in the weak* topology, and the weak* and uniform topologies coincide on $D_n(S; \beta H)$.

The proof of the proposition relies upon the result 3.2.3. and an elaboration of the convexity argument used in the introduction, basically the convexity of the function $t \rightarrow -t\log t$.

First we will show that the topologies coincide on $D_n(S; \beta H)$ by showing that for each n there is an m such that

$$D_n(S; \beta H) \subset D_m(H)$$

the coincidence then being a corollary of 3.2.3.

Now if $0 < \beta_0 < \beta$ we have for each $\omega_\rho \in D_n(S; \beta H)$

$$n \leq S(\beta H; \omega_\rho) = S(\beta_0 H; \omega_\rho) - (\beta - \beta_0)\hat{H}(\omega_\rho)$$

$$\leq \log Tr_{\mathcal{H}}(e^{-\beta_0 H}) - (\beta - \beta_0)\hat{H}(\omega_\rho)$$

where we have used the bound derived in the introduction. But then we conclude that

$$\hat{H}(\omega_\rho) \leq (\beta - \beta_0)^{-1} [\log Tr_{\mathcal{H}}(e^{-\beta_0 H}) - n]$$

thus $D_n(S; \beta H) \subset D_m(H)$ where

$$m = (\beta - \beta_0)^{-1} [\log Tr_{\mathcal{H}}(e^{-\beta_0 H}) - n] .$$

Next we deduce the upper semi-continuity of the conditional entropy over $E_{\mathfrak{A}}$ by showing that each of the sets $D_n(S; \beta H)$ is closed in the weak* topology. But as $D_n(S; \beta H)$ is contained in a weak* closed set $D_m(H)$ it then suffices to show that $\omega_\rho \in D_m(H) \rightarrow S(\beta H; \omega_\rho)$ is upper semi-continuous.

Using the notation of the introduction, i.e. $\lambda_n(\rho)$ and ψ_n^ρ denote the eigenvalues and eigenvectors of ρ, we have

$$S(\beta H; \omega_\rho) - \log Tr_{\mathcal{H}}(e^{-\beta H}) = \sum_{n \geq 1} x_n$$

where

$$X_n = -\lambda_n(\rho)\log\lambda_n(\rho) - \lambda_n(\rho)\left[\beta h(\psi_n^\rho) + \log \mathrm{Tr}_{\mathcal{H}}(e^{-\beta H})\right] + \lambda_n(\rho) - \frac{(\psi_n^\rho, e^{-\beta H}\psi_n^\rho)}{\mathrm{Tr}_{\mathcal{H}}(e^{-\beta H})}$$

(h is the quadratic form associated with H). However successively applying the two convexity inequalities

$$-\beta h(\psi_n^\rho) = \log e^{-\beta h(\psi_n^\rho)} \leq \log(\psi_n^\rho, e^{-\beta H}\psi_n^\rho)$$

and

$$-x(\log x - \log y) \leq y - x \qquad , x, y > 0$$

one finds immediately that $X_n \leq 0$. Thus

$$S(\beta H; \omega_\rho) - \log \mathrm{Tr}_{\mathcal{H}}(e^{-\beta H}) = \inf_N \sum_{n \leq N} X_n .$$

Hence if we can show that the function occurring in the infimum on the right hand side of this equation is upper semi-continuous in the weak* topology on $D_m(H)$ then we can conclude that $\omega_\rho \in D_m(H) \longrightarrow S(\beta H; \omega_\rho)$ is upper semi-continuous because it is then given as the lower envelope of a family of upper semi-continuous functions (cf. 3.1.5.). However the weak* topology and the uniform topology coincide on $D_m(H)$ by the preceding argument and hence we have to consider the continuity of the X_n in the uniform topology.

But the term

$$\omega_\rho \longrightarrow -\lambda_n(\rho)\log\lambda_n(\rho) - \lambda_n(\rho)\log \mathrm{Tr}_{\mathcal{H}}(e^{-\beta H}) + \lambda_n(\rho)$$

is continuous in the uniform topology as a result of 3.1.10. The remaining term in X_n can be written as

$$r_n = \inf_m \left[-\beta\lambda_n(\rho)(\psi_n^\rho, H_m\psi_n^\rho) - \frac{(\psi_n^\rho, e^{-\beta H}\psi_n^\rho)}{\mathrm{Tr}_{\mathcal{H}}(e^{-\beta H})}\right]$$

where H_m is the truncated form of H used in the definition of $\hat{H}$ (3.2.2.). Thus r_n is the lower envelope of a family of functions which are continuous in the uniform topology by 3.1.10., and hence r_n is upper semi-continuous and the semi-continuity of $S(\beta H; \cdot)$ is proved. (Note that 3.1.10. actually states continuity of the eigenprojectors of ρ and not the eigenvectors which are not uniquely defined. But by 3.1.10. there is a choice of the ψ_n^ρ which ensures the continuity).

3.2.6. We finally prove the convexity properties stated in 3.2.5. From the concavity of $t \to -t\log t$ one has

$$-(\psi, \rho\log\rho\, \psi) \leq -(\psi, \rho\psi)\log(\psi, \rho\psi) \qquad , \; \|\psi\| = 1 \; , \rho \in N \; .$$

Hence if $\mathcal{F}$ is a complete orthonormal basis of vectors in $\mathcal{H}$

$$-\mathrm{Tr}_{\mathcal{H}}(\rho\log\rho) = \inf_{\mathcal{F}} \sum_{\psi \in \mathcal{F}} -(\psi, \rho\psi)\log(\psi, \rho\psi)$$

(the equality is attained for $\mathcal{F} = \{\psi_n^\rho\}_{n \geq 1}$).

Thus for $0 < \alpha < 1$ and $\rho_1, \rho_2 \in N$ one has

$$\begin{aligned}
-\mathrm{Tr}_{\mathcal{H}}&((\alpha\rho_1 + (1-\alpha)\rho_2)\log(\alpha\rho_1 + (1-\alpha)\rho_2)) \\
&= \inf_{\mathcal{F}} \sum_{\psi \in \mathcal{F}} -\{\alpha(\psi, \rho_1\psi) + (1-\alpha)(\psi, \rho_2\psi)\}\log\{\alpha(\psi, \rho_1\psi) + (1-\alpha)(\psi, \rho_2\psi)\} \\
&\geq \inf_{\mathcal{F}} \sum_{\psi \in \mathcal{F}} -\alpha(\psi, \rho_1\psi)\log(\psi, \rho_1\psi) - (1-\alpha)(\psi, \rho_2\psi)\log(\psi, \rho_2\psi) \\
&\geq -\alpha\,\mathrm{Tr}_{\mathcal{H}}(\rho_1\log\rho_1) - (1-\alpha)\,\mathrm{Tr}_{\mathcal{H}}(\rho_2\log\rho_2)
\end{aligned}$$

i.e. the entropy is a concave function over N. But as $\hat{H}$ is affine one concludes immediately that

$$S(\beta H \,;\, \alpha\omega_{\rho_1} + (1-\alpha)\omega_{\rho_2}) \geq \alpha\, S(\beta H \,;\, \omega_{\rho_1}) + (1-\alpha)\, S(\beta H \,;\, \omega_{\rho_2})$$

for $0 < \alpha < 1$ and $\omega_{\rho_1}, \omega_{\rho_2} \in N$. But this inequality remains valid for all $\omega_1, \omega_2 \in E_{\mathfrak{A}}$ because if $\omega_1 \notin N$ then both sides take the value minus infinity.

The second property is more difficult to prove. It depends upon the following result ; let A and B be bounded self-adjoint operators such that

$$A \geq B > 0$$

then it follows that

$$\log A \geq \log B \; .$$

We will not prove this property (cf. exercise 4). Accepting this result the rest of the proof is straightforward. First one deduces that for $\rho_1, \rho_2 \in N$ and $0 < \alpha < 1$

$$\begin{aligned}
-\alpha \operatorname{Tr}_{\mathcal{H}}\left(\rho_1 \operatorname{Log}(\alpha\rho_1 + (1-\alpha)\rho_2)\right) &\leq -\alpha \operatorname{Tr}_{\mathcal{H}}\left(\rho_1 \operatorname{Log} \alpha\rho_1\right) \\
&= -\alpha \log\alpha - \alpha \operatorname{Tr}_{\mathcal{H}}\left(\rho_1 \operatorname{Log} \rho_1\right)
\end{aligned}$$

and similarly,

$$-(1-\alpha)\operatorname{Tr}_{\mathcal{H}}\left(\rho_2 \operatorname{Log}(\alpha\rho_1 + (1-\alpha)\rho_2)\right) \leq -(1-\alpha)\log(1-\alpha) - (1-\alpha)\operatorname{Tr}_{\mathcal{H}}\left(\rho_2 \log\rho_2\right) .$$

Adding these two inequalities and using the affinity of $\hat{H}$ one then finds

$$\begin{aligned}
S(\beta H ; \alpha\omega_{\rho_1} + (1-\alpha)\omega_{\rho_2}) &\leq \alpha S(\beta H ; \omega_{\rho_1}) + (1-\alpha) S(\beta H ; \omega_{\rho_2}) - \alpha\log\alpha - (1-\alpha)\operatorname{Log}(1-\alpha) \\
&\leq \alpha S(\beta H ; \omega_{\rho_1}) + (1-\alpha) S(\beta H ; \omega_{\rho_2}) + \log 2
\end{aligned}$$

for all $0 < \alpha < 1$ and $\omega_{\rho_1}, \omega_{\rho_2} \in N$. But again this relation remains valid for all $\omega_1, \omega_2 \in E_{\mathfrak{A}}$ because on the complement of N the conditional entropy has the value minus infinity.

BIBLIOGRAPHY

For the mathematical material we have discussed in section §1 consult

G. Köthe

Topologische Lineare Räume, - Springer-Verlag (Berlin) 1960.

J. Dixmier

Les C* algèbres et leurs représentations. - Gauthier-Villars (Paris) 1964.

F. Riesz and B. Nagy

Leçons d'analyse fonctionnelle - Gauthier-Villars (Paris) 1965.

The material of section §2 is extracted from

O. Lanford and D.W. Robinson

J. Math. Phys. 9 1120 (1968)

D.W. Robinson

Commun. Math. Phys. 19 219 (1970)

EXERCISES

1. Prove that the state ρ^* is the unique state for which the supremum is attained in the variational problem discussed in the introduction.

(Hint : If $(\lambda_m)_{m\geq 1}$ are the eigenvalues of H and ψ_m an associated orthonormal basis of eigenfunctions then

$$S_\Lambda(\beta;\rho) - P_\Lambda(\beta) = V(\Lambda)^{-1}\sum_{n\geq 1}\{-(\psi_n,\rho\log\rho\,\psi_n) + (\psi_n,\rho\,\psi_n)(\log e^{-\beta\lambda_n})/\mathrm{Tr}(e^{-\beta H})\}$$

$$\leq V(\Lambda)^{-1}\sum_{n\geq 1} -(\psi_n,\rho\,\psi_n)\left[\log(\psi_n,\rho\,\psi_n) - (\log e^{-\beta\lambda_n})/\mathrm{Tr}(e^{-\beta H})\right]$$

$$\leq 0$$

where both inequalities use the convexity of $t \to -t\log t$. Deduce that the first is a strict equality if, and only if, ψ_n is an eigenfunction of ρ and the second if, and only if, the corresponding eigenvalue is 0 or $e^{-\beta\lambda_n}/\mathrm{Tr}(e^{-\beta H})$ Use the normalization of ρ to rule out the first possibility).

2. Prove the continuity of the eigenprojectors stated in 3.1.10.

(Hint : Refer to pages 365 - 366 of Riesz and Nagy, Leçons d'analyse fonctionnelle).

3. Let A and B be positive self-adjoint operators with associated forms a and b . Prove that if $A \geq B > 0$ then

$$B^{-1} \geq A^{-1} .$$

(Hint : For $\psi \in \mathcal{H}$ set $\psi_a = A^{-1}\psi$, $\psi_b = B^{-1}\psi$ then

$$(A^{-1}\psi, \psi)^2 = (\psi_a, B\psi_b)^2 \leq b(\psi_a)\, b(\psi_b)$$

$$\leq a(\psi_a)\, b(\psi_b) = (A^{-1}\psi,\psi)(B^{-1}\psi,\psi) . \quad)$$

4. Under the same conditions as exercise 3. prove that

$$\log A \geq \log B$$

(Hint : Use the integral representation

$$\log A - \log B = \int_0^\infty dz \left\{ \frac{1}{B+z} - \frac{1}{A+z} \right\} . \quad)$$

CHAPTER IV

THE MAXIMUM ENTROPY PRINCIPLE

INTRODUCTION

In this final chapter we complete the discussion begun in Chapter 3 by showing that the thermodynamic pressure is given by the maximum of the conditional entropy per unit volume where the maximum is taken over a suitably chosen set of translationally invariant states. We also derive some properties of the states for which the maximum is attained, the equilibrium states of the thermodynamic system.

We begin in section §1 by giving a more precise characterization of the states which enter into the variational principle and we show that these states can be embedded in the set of translationally invariant states over a suitably chosen C* algebra $\mathfrak{A}$. This latter set is compact in the weak* topology of $\mathfrak{A}$ and in section §2 , we prove that the mean conditional entropy is an affine upper semi-continuous function over the set. In section §3 we demonstrate that the supremum of the mean conditional entropy gives the thermodynamic pressure for the set of finite range interactions discussed in section §3 of chapter 2. We conclude, in section §4 , with a brief discussion of the equilibrium states.

§1 - THE THERMODYNAMIC STATES

4.1.1. Each finite system of particles is described with the aid of the Hilbert spaces $\mathcal{H}(\Lambda)$ of 1.3.2. In 1.3.3. we have shown that if Λ_1 and Λ_2 are disjoint subsets of R^ν then

$$\mathcal{H}(\Lambda_1 \cup \Lambda_2) = \mathcal{H}(\Lambda_1) \otimes \mathcal{H}(\Lambda_2) .$$

We consider the thermodynamic or infinite, system of particles to be describable in terms of the family of all its finite subsystems, i.e. characterizable with the aid of the family $\{\mathcal{H}(\Lambda) ; \Lambda \subset R^\nu\}$ of Hilbert spaces.

4.1.2. A quantum-mechanical state of the finite system on Λ is given by a density matrix ρ_Λ on $\mathcal{H}(\Lambda)$ and a thermodynamic state will be more generally characterized by a family $\{\rho_\Lambda ; \Lambda \subset R^\nu\}$ of compatible density matrices. The condition of compatibility being defined by

$$\rho_{\Lambda_2} = \mathrm{Tr}_{\mathcal{H}(\Lambda_1)}(\rho_{\Lambda_1 \cup \Lambda_2})$$

whenever

$$\Lambda_1 \cap \Lambda = \phi .$$

This condition can be understood in the following manner. If A is a bounded operator on $\mathcal{H}(\Lambda_1)$ then it can be identified with the operator $A \otimes \mathbb{1}_{\Lambda_2}$ on $\mathcal{H}(\Lambda_1 \cup \Lambda_2)$. The expectation valued of this operator in the state $\{\rho_\Lambda\}$ is then given by **two** expressions which should for consistency be equal namely

$$\mathrm{Tr}_{\mathcal{H}(\Lambda_1)}(\rho_{\Lambda_1} A) = \mathrm{Tr}_{\mathcal{H}(\Lambda_1 \cup \Lambda_2)}(\rho_{\Lambda_1 \cup \Lambda_2} A \otimes \mathbb{1}_{\Lambda_2}) .$$

Now if this condition is required for a complete set of observables A , or in mathematical language an irreducible algebra of operators on $\mathcal{H}(\Lambda_1)$, the density matrices must be related by the foregoing condition.

4.1.3. For each $\Lambda \subset R^\nu$ and $a \in R^\nu$ we can define a unitary mapping $U_{\Lambda,a}$ of $\mathcal{H}(\Lambda)$ into $\mathcal{H}(\Lambda + a)$ by

$$(U_{\Lambda,a}\psi)(X) = \psi(X-a)$$

for all

$$\psi \in \mathcal{H}(\Lambda) .$$

The state $\{\rho_\Lambda\}$ will be defined to be translationnally invariant if

$$\rho_{\Lambda+a} = U_{\Lambda,a}\,\rho_\Lambda\,U_{\Lambda,a}^{-1}$$

for all regions Λ and translations a. If on the other hand this last relation is valid for all Λ but only for a taking values in a lattice Z^ν then we describe $\{\rho_\Lambda\}$ as <u>periodic</u>.

The translationally invariant states $\{\rho_\Lambda\}$ will be the set of states figuring in the variational principle which we derive in the sequel. Although, this set can be directly topologised in the manner of 4.1.7. it will be convenient for our purposes to embed it in a larger set which is compact in a suitable topology. This embedding and topological characterization will be achieved with the aid of a C* algebra with local structure.

4.1.4. Let $\{\mathfrak{A}_\Lambda\}$ be a family of C* algebras satisfying the following conditions.

1. For each Λ, $\mathfrak{A}_\Lambda$ is an irreducible C* algebra of operators on the Hilbert space $\mathcal{H}(\Lambda)$ which contains the identity operator $\mathbb{1}_\Lambda$.

2. If $\Lambda_1 \cap \Lambda_2 = \phi$ then $\mathfrak{A}_{\Lambda_1} \subset \mathfrak{A}_{\Lambda_1 \cup \Lambda_2}$ where we implicitly identify $\mathfrak{A}_{\Lambda_1}$ on $\mathcal{H}(\Lambda_1)$ with the algebra $\mathfrak{A}_{\Lambda_1} \otimes \mathbb{1}_{\Lambda_2}$ on $\mathcal{H}(\Lambda_1 \cup \Lambda_2)$.

3. The mapping $A \in \mathcal{L}(\mathcal{H}(\Lambda)) \longrightarrow \tau_a A \in \mathcal{L}(\mathcal{H}(\Lambda + a))$ defined by

$$\tau_a A = U_{\Lambda,a}\,A\,U_{\Lambda,a}^{-1}$$

is an automorphism of $\mathfrak{A}_\Lambda$ into $\mathfrak{A}_{\Lambda+a}$ for all regions Λ and translates a.

Now using condition 2 we can define the C* algebra $\mathfrak{A}$ to be the unique minimal completion of the $\{\mathfrak{A}_\Lambda\}$ in the uniform topology. Explicitly the family $\{\mathfrak{A}_\Lambda\}$ form an incomplete metric space with metric distance $\|A_1 - A_2\|$ for $A_1 \in \mathfrak{A}_{\Lambda_1}$, $A_2 \in \mathfrak{A}_{\Lambda_2}$ where $\|\cdot\|$ denotes the norm on $\mathcal{H}(\Lambda_1 \cup \Lambda_2)$ Such a space has a unique minimum completion which is a Banach space $\mathfrak{A}$. But the family $\{\mathfrak{A}_\Lambda\}$ also constitutes a * algebra and hence $\mathfrak{A}$ can be interpreted as a C* algebra with the algebraic rules obtained from the $\{\mathfrak{A}_\Lambda\}$ and by continuity. Note that the automorphism group $A \in \mathfrak{A}_\Lambda \to \tau_a A \in \mathfrak{A}_{\Lambda+a}$ also extends by continuity to a group of automorphisms of $\mathfrak{A}$.

<u>4.1.5.</u> Each state $\{\rho_\Lambda\}$ in the sense of 4.1.3. determines a state of $\mathfrak{A}$ by the definition

$$\omega_\rho(A) = \mathrm{Tr}_{\mathcal{H}(\Lambda)}(\rho_\Lambda A)$$

for all $A \in \mathfrak{A}_\Lambda$ and all Λ and through extension by continuity to all other $A \in \mathfrak{A}$. The set of states of $\mathfrak{A}$ obtained in this manner will be called <u>locally normal states</u> and will be denoted by L. The set L is a proper subset of $E_\mathfrak{A}$. We define $E_\mathfrak{A}(R^\nu)$ to be the set of <u>translationally invariant</u> states over $\mathfrak{A}$, i.e. the set of states $\omega \in E_\mathfrak{A}$ satisfying

$$\omega(A) = \omega(\tau_a A) \qquad , \quad a \in R^\nu \quad , \quad A \in \mathfrak{A} \, .$$

If $\{\rho_\Lambda\}$ is a translationally invariant state then the associated locally normal state is a translationally invariant state over $\mathfrak{A}$. We denote the set of such states by $L(R^\nu)$.

Similarly we could define the set of periodic states over $\mathfrak{A}$ and then each periodic $\{\rho_\Lambda\}$ would determine a periodic state.

As we have assumed that each $\mathfrak{A}_\Lambda$ contains the identity operator $\mathbb{1}_\Lambda$ the algebra $\mathfrak{A}$ must contain an identity and hence $E_\mathfrak{A}$ is weak* compact by 3.1.6. Note further that the set $E_\mathfrak{A}(R^\nu)$ of translationally invariant states is convex and compact in the weak* topology. The convexity is obvious and the compactness follows from noting that $E_\mathfrak{A}(R^\nu)$ is weak* closed because if ω_α is a directed set of translationally invariant states which converges weak* to $\omega \in E_\mathfrak{A}$ then ω is clearly invariant.

The set $E_\mathfrak{A}(R^\nu)$ is the set which will be used to construct the variational principle.

<u>4.1.7.</u> The set of states $E_\mathfrak{A}$, or any of its subsets, can be equipped not only with the weak* topology of $\mathfrak{A}$ but also the uniform topology. The special structure of $\mathfrak{A}$ allows us however to introduce a third topology on $E_\mathfrak{A}$ which is intermediate to these two topologies. This third topology will be called the <u>locally uniform topology</u> and is specified by the set of neighbourhoods

$$\mathcal{V}(\omega; \Lambda, \epsilon) = \{ \omega' \, ; \, \omega' \in E_\mathfrak{A} \ \sup_{A \in \mathfrak{A}_\Lambda} |\omega'(A) - \omega(A)| / \|A\| < \epsilon \}$$

where $\omega \in E_\mathfrak{A}$, $\Lambda \subset R^\nu$ and $\epsilon > 0$.

Clearly the locally uniform topology is finer than the weak* topology and coarser than the uniform topology. It is a metric topology and one can construct a metric for it as follows. Let Λ_n denote balls of radius $n = 1, 2, \ldots$ centred at the origin and introduce the definition

$$\| \omega_1 - \omega_2 \|_n = \sup_{A \in \mathfrak{A}_{\Lambda_n}} | \omega_1(A) - \omega_2(A) | / \|A\| \quad , \ \omega_1, \omega_2 \in E_{\mathfrak{A}}.$$

Then by a standard argument the locally uniform topology is determined by the metric

$$\| \omega_1 - \omega_2 \| = \sum_{n \geq 1} \frac{1}{2^n} \frac{\| \omega_1 - \omega_2 \|_n}{1 + \| \omega_1 - \omega_2 \|_n} .$$

The normal states over $\mathfrak{A}$ are closed and complete in the uniform topology (3.1.7.) and similarly the locally normal states L are closed and complete in the locally uniform topology. The same properties are of course true for the translationally invariant locally normal states $L(R^\nu)$.

4.1.8. In the above discussion the C* algebra $\mathfrak{A}$ was fixed by demanding that its generating family $\{\mathfrak{A}_\Lambda\}$ satisfy a certain number of structural relations (4.1.4.) Note that these relations are all satisfied if we identify each $\mathfrak{A}_\Lambda$ with the algebra $\mathcal{L}(\mathcal{H}(\Lambda))$ of all bounded operators on $\mathcal{H}(\Lambda)$. Thus a C* algebra with the required properties certainly exists but is by no means unique. We will not construct other examples but they can be obtained from discussing the canonical commutation relations in the Weyl form. This arbitrariness is however irrelevant for the results which we discuss in the following sections.

4.1.9. In fact the states $\{\rho_\Lambda\}$ (or more specifically the translationally invariant states) can be directly considered as a topological space L (or $L(R^\nu)$) equipped with the locally uniform topology. The neighbourhoods of $\omega = \{\rho_\Lambda\}$ are then

$$\mathcal{U}(\omega ; \Lambda, \epsilon) = \{ \omega' ; \ \omega' = \{\rho'_\Lambda\} , \ |||\rho'_\Lambda - \rho_\Lambda||| < \epsilon \}$$

where $||| \cdot |||$ indicates the trace norm introduced in 3.1.9. With this topology L is a complete metric space and $L(R^\nu)$ is a closed, and hence compelte, subspace.

§2 - THE MEAN CONDITIONAL ENTROPY

4.2.1. In this section we show that for each of the Hamiltonians of chapter 2 with elastic boundary conditions one can extend the arguments of chapter 3 to define the energy and conditional entropy per unit volume as affine semi-continuous functions over the invariant states $E_{\mathfrak{A}}(R^\nu)$.

We will take $\{H_\Lambda\}$ to be a family of self-adjoint operators on the Hilbert spaces $\{\mathcal{H}(\Lambda)\}$ and in particular we take the following identification

$$H_\Lambda = K^\mu_{0,\Lambda} \dotplus U_\Lambda \quad , \quad \mu < 0$$

where $K^\mu_{0,\Lambda}$ is the (kinetic energy) operator of 1.3.4. and U_Λ is an interaction operator satisfying the conditions of 2.2.2. and 2.2.3. Thus each H_Λ is positive and $\exp(-\beta H_\Lambda)$ is of trace class (on $\mathcal{H}(\Lambda)$) for $\beta > 0$ as a consequence of the estimates of 2.1.2. and 2.2.5. Further we have for $\Lambda_1 \cap \Lambda_2 = \emptyset$ (and the $\partial\Lambda$ sufficiently smooth)

$$H_{\Lambda_1 \cup \Lambda_2} \geqslant (H_{\Lambda_1} \otimes 1\!\!1_{\Lambda_2})^{**} \dotplus (1\!\!1_{\Lambda_1} \otimes H_{\Lambda_2})^{**} \qquad (*)$$

$$= (H_{\Lambda_1} \otimes 1\!\!1_{\Lambda_2} + 1\!\!1_{\Lambda_1} \otimes H_{\Lambda_2})^{**}$$

and

$$H_{\Lambda+a} = U_{\Lambda,a} H_\Lambda U_{\Lambda,a}^{-1}$$

(The inequality can be deduced from 2.2.8. ; cf. also exercise 3 of chapter 2).

Now for each H_Λ we can repeat the construction of §2 of chapter 3 to define a family $\{\hat{H}_\Lambda\}$ of functionals $\omega \in E_{\mathfrak{A}_\Lambda} \to \hat{H}_\Lambda(\omega)$ representing the expectation values of the local Hamiltonians H_Λ. We consider next the properties of this family as Λ varies using the property $(*)$.

Proposition 4.2.2. Let $\omega \in E_{\mathfrak{A}}(R^\nu)$ be a translationally invariant state over $\mathfrak{A}$ then the function $\Lambda \subset R^\nu \to \hat{H}_\Lambda(\omega) \in [0, +\infty]$ is super-additive

$$\hat{H}_{\Lambda_1 \cup \Lambda_2}(\omega) \geqslant \hat{H}_{\Lambda_1}(\omega) + \hat{H}_{\Lambda_2}(\omega) \qquad \Lambda_1 \cap \Lambda_2 = \emptyset$$

and invariant

$$\hat{H}_\Lambda(\omega) = \hat{H}_{\Lambda+a}(\omega) \qquad \Lambda \subset R^\nu,\ a \in R^\nu .$$

It follows that the limit

$$\hat{H}(\omega) = \lim_{L^{(1)},\ldots,L^{(\nu)}\to\infty} \hat{H}_\Lambda(\omega)/V(\Lambda)$$

exists and

$$\hat{H}(\omega) = \sup_{L^{(1)},\ldots,L^{(\nu)}} \hat{H}_\Lambda(\omega)/V(\Lambda) \;.$$

Thus if $\hat{H}(\omega)$ is finite then $\omega \in L(R^\nu)$. The function

$$\omega \in E_{\mathfrak{A}}(R^\nu) \longrightarrow \hat{H}(\omega) \in [0, +\infty]$$

is affine and lower semi-continuous in the weak*topology of $\mathfrak{A}$.

The proof of the super-additivity property is straightforward. If ω is not locally normal, both sides of the inequality take the value $+\infty$. If on the other hand ω is determined by the family $\{\rho_\Lambda\}$ then

$$\hat{H}_{\Lambda_1\cup\Lambda_2}(\omega) = \sum_{m\geq 1} \lambda_m(\rho_{\Lambda_1\cup\Lambda_2})\, h_{\Lambda_1\cup\Lambda_2}(\psi_m^\rho)$$

$$\geq \sum_{m\geq 1} \lambda_m(\rho_{\Lambda_1\cup\Lambda_2})\left[(h_{\Lambda_1}\otimes \mathbb{1}_{\Lambda_2})(\psi_m^\rho) + (\mathbb{1}_{\Lambda_1}\otimes h_{\Lambda_2})(\psi_m^\rho)\right]$$

where we have used an obvious notation for the forms associated with $H_{\Lambda_1}\otimes \mathbb{1}_{\Lambda_2}$ etc. But then introducing the spectral decompositions of H_{Λ_1} and H_{Λ_2} we have

$$\hat{H}_{\Lambda_1\cup\Lambda_2}(\omega) \geq \sum_{m,n,p} \lambda_m(\rho_{\Lambda_1\cup\Lambda_2})\left[\lambda_n(H_{\Lambda_1}) + \lambda_p(H_{\Lambda_2})\right] |(\psi_m^\rho, \psi_n^{H_{\Lambda_1}}\otimes \psi_p^{H_{\Lambda_2}})|^2$$

$$= \sum_n \lambda_n(H_{\Lambda_1})(\psi_n^{H_{\Lambda_1}}, \rho_{\Lambda_1}\psi_n^{H_{\Lambda_1}}) + \lambda_n(H_{\Lambda_2})(\psi_n^{H_{\Lambda_2}}, \rho_{\Lambda_2}\psi_n^{H_{\Lambda_2}})$$

$$= \hat{H}_{\Lambda_1}(\omega) + \hat{H}_{\Lambda_2}(\omega)$$

where the positivity of all terms has allowed us to interchange the orders of summation and we have used the compatibility condition of 4.1.2.

The invariance property follows straightforwardly from the invariance of the Hamiltonians and the invariance of the state ω. We omit the formal details.

The existence of the limit and its expression as a supremum follows by the same lemma used in 2.1.8. and 2.1.9. (cf. exercise 1 of chapter 2). As $\hat{H}(\omega) < +\infty$ implies that $\hat{H}_\Lambda(\omega) < +\infty$ for each parallelepiped Λ it follows that ω is locally normal.

Finally $\hat{H}$ is affine because it is the limit of a family of affine functions (3.2.3.) and lower semi-continuous because it is the upper envelope of a family of lower semi-continuous functions (3.2.3. and 3.1.5.).

Next consider the conditional entropy defined in the manner of 3.2.4. but using the family of Hamiltonians $\{\mathcal{H}_\Lambda\}$ of 4.2.1. Thus we have a family of conditional entropies $\omega \in E_{\mathfrak{A}_\Lambda} \longrightarrow S_\Lambda(\beta H;\omega)$.

Proposition 4.2.3. Let $\omega \in E_{\mathfrak{A}}(R^\nu)$ be a translationally invariant state over $\mathfrak{A}$ then the function $\Lambda \subset R^\nu \longrightarrow S_\Lambda(\beta H;\omega) \in [-\infty, \log Tr_{\mathcal{H}(\Lambda)}(e^{-\beta H_\Lambda})]$ is sub-additive

$$S_{\Lambda_1\cup\Lambda_2}(\beta H;\omega) \leq S_{\Lambda_1}(\beta H;\omega) + S_{\Lambda_2}(\beta H;\omega) \quad , \quad \Lambda_1\cap\Lambda_2 = \phi$$

and invariant

$$S_\Lambda(\beta H;\omega) = S_{\Lambda+a}(\beta H;\omega) \quad , \quad \Lambda\subset R^\nu, a\in R^\nu .$$

It follows that the limit

$$S(\beta H;\omega) = \lim_{L^{(1)},\dots,L^{(\nu)}\to\infty} S_\Lambda(\beta H;\omega)/V(\Lambda)$$

exists and

$$S(\beta H;\omega) = \inf_{L^{(1)},\dots,L^{(\nu)}} S_\Lambda(\beta H;\omega)/V(\Lambda) .$$

The function $\omega \in E_{\mathfrak{A}}(R^\nu) \longrightarrow S(\beta H;\omega) \in [-\infty, P(\beta,\mu)]$ is affine and upper semi-continuous in the weak* topology.

First note that the upper bound for $S_\Lambda(\beta H;\omega)$ was derived in introduction to chapter 3. With the identification $H_\Lambda = K^\mu_{0,\Lambda} \dotplus U_\Lambda$ the pressure

$$P(\beta,\mu) = \lim_{L^{(1)},\dots,L^{(\nu)}\to\infty} \frac{1}{V(\Lambda)} \log Tr_{\mathcal{H}(\Lambda)}(e^{-\beta H_\Lambda})$$

is proved to exist in 2.2.9. and clearly constitutes an upper bound for $S(\beta H;\omega)$ when this latter quantity exists.

Now to prove the subadditivity of $S_\Lambda(\beta H;\omega)$ it essentially suffices to consider the case that ω is locally normal because in the other case both sides of the inequality are equal to $-\infty$. But as the conditional entropy of a state $\{\rho_\Lambda\}$ is defined as the entropy minus a multiple of the energy and the latter function is super-additive by 4.2.2. it suffices to show that

$$-Tr_{\mathcal{H}(\Lambda_1\cup\Lambda_2)}(\rho_{\Lambda_1\cup\Lambda_2}\log\rho_{\Lambda_1\cup\Lambda_2}) \leq -Tr_{\mathcal{H}(\Lambda_1)}(\rho_{\Lambda_1}\log\rho_{\Lambda_1}) - Tr_{\mathcal{H}(\Lambda_2)}(\rho_{\Lambda_2}\log\rho_{\Lambda_2})$$

whenever $\Lambda_1 \cap \Lambda_2 = \emptyset$. This can however be proved by a convexity argument using the compatibility condition of 4.1.2. as follows. First we have

$$-\lambda_m(\rho_{\Lambda_1\cup\Lambda_2})\left[\log\lambda_m(\rho_{\Lambda_1\cup\Lambda_2}) - \log(\psi_m^\rho, \rho_{\Lambda_1}\otimes\rho_{\Lambda_2}\psi_m^\rho)\right] \leq (\psi_m^\rho, \rho_{\Lambda_1}\otimes\rho_{\Lambda_2}\psi_m^\rho) - \lambda_m(\rho_{\Lambda_1\cup\Lambda_2})$$

by the convexity inequality used in the introduction of chapter 3. and 3.2.5. Thus summing over m and using the normalization of the density matrices one has

$$-\mathrm{Tr}_{\mathcal{H}(\Lambda_1\cup\Lambda_2)}(\rho_{\Lambda_1\cup\Lambda_2}\log\rho_{\Lambda_1\cup\Lambda_2}) \leq -\sum_{m\geq 1}\lambda_m(\rho_{\Lambda_1\cup\Lambda_2})\log(\psi_m^\rho, \rho_{\Lambda_1}\otimes\rho_{\Lambda_2}\psi_m^\rho)$$
$$= -\sum_{m\geq 1}\lambda_m(\rho_{\Lambda_1\cup\Lambda_2})\log\sum_{n,p\geq 1}\lambda_n(\rho_{\Lambda_1})\lambda_p(\rho_{\Lambda_2})|(\psi_m^\rho, \psi_n^\rho\otimes\psi_p^\rho)|^2$$

where if $(\psi_m^\rho, \rho_{\Lambda_1}\otimes\rho_{\Lambda_2}\psi_m^\rho) = 0$ and $\lambda_m(\rho_{\Lambda_1\cup\Lambda_2}) = 0$ we adopt the convention that the corresponding term is zero. Now using the convexity of the logarithm we find

$$-\mathrm{Tr}_{\mathcal{H}(\Lambda_1\cup\Lambda_2)}(\rho_{\Lambda_1\cup\Lambda_2}\log\rho_{\Lambda_1\cup\Lambda_2}) \leq -\sum_{m,n,p\geq 1}\lambda_m(\rho_{\Lambda_1\cup\Lambda_2})|(\psi_m^\rho, \psi_n^\rho\otimes\psi_p^\rho)|^2 \cdot$$
$$\cdot\left[\log\lambda_n(\rho_{\Lambda_1}) + \log\lambda_p(\rho_{\Lambda_2})\right]$$
$$= -\mathrm{Tr}_{\mathcal{H}(\Lambda_1)}(\rho_{\Lambda_1}\log\rho_{\Lambda_1}) - \mathrm{Tr}_{\mathcal{H}(\Lambda_2)}(\rho_{\Lambda_2}\log\rho_{\Lambda_2})$$

where the last step uses 4.1.2.

The invariance of S_Λ is again straightforward and the existence of $S(\beta H;\omega)$ and its equality with the infimum is again a consequence of the sub-additivity, invariance and the uniform upper boundedness of $S_\Lambda(\beta H;\omega)/V(\Lambda)$ (for the last point cf. 2.2.5. and 2.1.2.).

Now although $\omega\in E_{\mathfrak{A}_\Lambda} \to S_\Lambda(\beta H;\omega)$ is not affine we have from 3.2.5. that

$$\frac{S_\Lambda(\beta H;\alpha\omega_1+(1-\alpha)\omega_2)}{V(\Lambda)} \geq \alpha\frac{S_\Lambda(\beta H;\omega_1)}{V(\Lambda)} + (1-\alpha)\frac{S_\Lambda(\beta H;\omega_2)}{V(\Lambda)}$$

$$\frac{S_\Lambda(\beta H;\alpha\omega_1+(1-\alpha)\omega_2)}{V(\Lambda)} \leq \alpha\frac{S_\Lambda(\beta H;\omega_1)}{V(\Lambda)} + (1-\alpha)\frac{S_\Lambda(\beta H;\omega_2)}{V(\Lambda)} + \frac{\log 2}{V(\Lambda)}$$

$0<\alpha<1$, $\omega_1,\omega_2\in E_{\mathfrak{A}_\Lambda}$. Thus in the limit $\omega\to S(\beta H;\omega)$ is affine. This latter function is then upper semi-continuous because it is given by the proposition and 3.2.5. as the lower envelope of a family of upper semi-continuous functions.

4.2.4. At this point we have established that the conditional entropy is an upper semi-continuous function over the compact set of states $E_{\mathfrak{A}}(R^\nu)$ (both statements with reference to the weak* topology of $\mathfrak{A}$). These are the essential points that will be used to obtain the variational principle of the following section but to actually show that the thermodynamic pressure $P(\beta,\mu)$ is reached as the largest value of $\{S(\beta H;\omega)\ ;\ \omega\in E_{\mathfrak{A}}(R^\nu)\}$ we will need to assume that we have a finite range interaction of the type discussed in chapter 2 section §3 . We will then construct invariant states as averages of periodic states and as the limits of such averages. We will next briefly discuss this procedure and properties of the conditional entropy in relation to the periodic states.

4.2.5. Take $b^{(1)},\ldots,b^{(\nu)}>0$ and let $E_{\mathfrak{A}}(Z^\nu)$ denote the class of states over $\mathfrak{A}$ which are periodic in the sense that

$$\omega(A)=\omega(\tau_x A)$$

for all $A\in\mathfrak{A}$ and all x of the form $(n^{(1)}b^{(1)},\ldots,n^{(\nu)}b^{(\nu)})$ where $n^{(1)},\ldots,n^{(\nu)}$ are integers.

If ω is a periodic state then it follows that each of the states $\tau_x\omega$ defined by

$$(\tau_x\omega)(A)=\omega(\tau_x A)$$

with $0<x^{(1)}<b^{(1)},\ldots,0<x^{(\nu)}<b^{(\nu)}$ are also periodic.

If further the function $x\to\tau_x\omega$ is integrable then the state $\widetilde{\omega}$ defined by

$$\widetilde{\omega}(A)=\frac{1}{b^{(1)}b^{(2)}\ldots b^{(\nu)}}\int_{0<x^{(i)}\leq b^{(i)}}dx\ (\tau_x\omega)(A)$$

is an invariant state, i.e. $\widetilde{\omega}\in E_{\mathfrak{A}}(R^\nu)$.

[As we have not assumed any continuity of the group of automorphisms $A\in\mathfrak{A}\longrightarrow\tau_x A\in\mathfrak{A}$ the function $x\to\tau_x\omega$ is not necessarily continuous. However if ω is a locally normal state then this function is certainly continuous for the following reason. First note that as the family $\{\mathfrak{A}_\Lambda\}$ is uniformly dense in $\mathfrak{A}$ it suffices to prove that $x\to(\tau_x\omega)(A)$ is continuous for each A in some $\mathfrak{A}_\Lambda$. But as ω is locally normal this a question of the continuity of

$$x \to \mathrm{Tr}_{\mathcal{H}(\Lambda')}\left(\rho_{\Lambda'} U_{\Lambda,x} A U_{\Lambda,x}^{-1}\right)$$

where Λ' is a region which is large enough to contain $\Lambda + x$ for all the values of x considered. Thus the continuity property in this case rests basically on the continuity properties of the unitaries $U_{\Lambda,x}$ and can be explicitly verified. We will omit the details of this verification. (cf. exercise 3)]

4.2.6. If now we replace the set of invariant states $E_{\mathfrak{A}}(R^\nu)$ by the set of periodic states in the discussion of 4.2.2. and 4.2.3. then very little is changed. The conditions of invariance for $\hat{H}_\Lambda$ and S_Λ are replaced by conditions of periodicity but the sub-additivity inequalities etc. are unchanged. It is then easy to see the mean energy $\hat{H}$ and the mean conditional entropy S can be defined as functions over the periodic states and that these functions remain affine and semi-continuous. But note also that if ω is a periodic state then

$$S(\beta H ; \omega) = S(\beta H ; \tau_x \omega)$$

for all $x \in R^\nu$; this is checked using the sub-additivity and upper-boundedness of $\Lambda \to S_\Lambda$ in a straightforward manner which we will not detail. However if ω is a periodic locally normal state and $\tilde{\omega}$ is the invariant state constructed in the manner of 4.2.5. from ω it further follows that

$$S(\beta H ; \tilde{\omega}) = S(\beta H ; \omega) .$$

Clearly this would be a consequence of

$$S(\beta H ; \tilde{\omega}) = \frac{1}{b^{(1)} \cdots b^{(\nu)}} \int_{0 < x^{(i)} < b^{(i)}} dx \; S(\beta H ; \tau_x \omega) .$$

But this latter relation follows because S is affine and upper semi-continuous. (Inserting the explicit definition of $\tilde{\omega}$ the relation is essentially the relation of affinity but with the slight difference that the state $\tilde{\omega}$ is given by an integral combination of periodic states ; it is because of this some continuity is necessary to supplement the affinity). As the proof of this last statement needs a slightly complex approximation procedure, in which $\tilde{\omega}$ is approached by finite convex combinations of states, we will omit the proof and present the result as a "fait accompli".

§3 - THE THERMODYNAMIC VARIATIONAL PRINCIPLE

4.3.1. We are now in a position to derive the variational principle which has been the aim of the last two chapters. We will consider a slightly more restrictive class of Hamiltonians than we have used up to now ; we consider only the Hamiltonians in which the interaction is given by a positive, decreasing, potential ϕ of finite range. (cf. the definitions and discussion §3 of chapter 2). We will display the dependence of the Hamiltonians on the potentials by replacing our previous notation $H_\Lambda, \hat{H}_\Lambda, \hat{H}$ by $H^\phi_\Lambda, \hat{H}^\phi_\Lambda, \hat{H}^\phi$ etc. Although our notation apparently suppresses the dependence of H_Λ etc. on μ this can be restored by adopting the convention $\mu = \phi(\{x\})$.

Recall that the locally normal translationally invariant states over the C* algebra $\mathfrak{A}$ were denoted by $L(R^\nu)$. Now define $D(H^\phi)$ to be the set of states

$$D(H^\phi) = \{ \omega \, ; \, \omega \in E_{\mathfrak{A}}(R^\nu) \, , \, \hat{H}^\phi(\omega) < +\infty \}$$

and note that it follows from our construction 4.2.2. that $D(H^\phi) \subset L(R^\nu)$.

If $\omega \in D(H^\phi)$ we can unambiguously define the mean entropy $\omega \in D(H^\phi) \to S(\omega)$ by

$$S(\omega) = S(\beta H^\phi; \omega) + \beta \hat{H}^\phi(\omega)$$

and we then have

$$S(\omega_\rho) = \lim_{\Lambda \to \infty} - \frac{1}{V(\Lambda)} \mathrm{Tr}_{\mathcal{H}(\Lambda)}(\rho_\Lambda \log \rho_\Lambda) \, .$$

This definition is actually unnecessary but we introduce it to emphasize the physical content of the following variational principle as the maximum of the mean entropy, or disorder, at fixed energy density.

<u>Proposition 4.3.2.</u> <u>If $P(\beta, \phi)$ denotes the thermodynamic pressure defined in section 2 of chapter 2 with the Hamiltonians $\{H^\phi_\Lambda\}$ then</u>

$$P(\beta, \phi) = \sup_{\omega \in D(H^\phi)} \left[S(\omega) - \beta \hat{H}^\phi(\omega) \right]$$

$$= \sup_{\omega \in E_{\mathfrak{A}}(R^\nu)} \left[S(\omega) - \beta \hat{H}^\phi(\omega) \right] \, .$$

First remark that we have from 4.2.3. that

$$S(\beta H^{\phi}\,;\,\omega) \leqslant P(\beta,\phi)$$

for all $\omega \in E_{\Omega}(R^{\nu})$. Thus to deduce the above result it suffices to construct an invariant state ω_{ϕ} such that

$$S(\beta H^{\phi}\,;\,\omega_{\phi}) = P(\beta,\phi)$$

and then to prove that $\omega_{\phi} \in D(H^{\phi})$. However once the first step is achieved the second results by noting that for $0 < \beta' < \beta$

$$\begin{aligned} P(\beta,\phi) = S(\beta H^{\phi}\,;\,\omega_{\phi}) &= S(\beta' H^{\phi}\,;\,\omega_{\phi}) - (\beta-\beta')\,H^{\phi}(\omega_{\phi}) \\ &\leqslant P(\beta'\,;\,\phi) - (\beta-\beta')\,H^{\phi}(\omega_{\phi}) \end{aligned}$$

i.e. we have

$$0 \leqslant H^{\phi}(\omega_{\phi}) \leqslant (\beta-\beta')^{-1}\left[P(\beta',\phi) - P(\beta,\phi)\right] < +\infty$$

We will now indicate the construction of a state ω_{ϕ} with the desired property.

Firstly the information on periodic states, given in 4.2.5. and 4.2.6. can be used to simplify the construction. In fact the construction of ω_{ϕ} can be reduced to the following problem.

Given $\epsilon > 0$ find a state ω^{ϵ}_{ϕ} which is periodic, with period $L^{(1)}, L^{(2)}, \ldots, L^{(\nu)}$, dependent on ϵ such that

$$S(\beta H^{\phi}\,;\,\omega^{\epsilon}_{\phi}) \geqslant C(\epsilon)\left[P(\beta,\phi) - \epsilon\right]$$

where $C(\epsilon) \to 1$ as $\epsilon \to 0$.

Once this problem is solved the construction of ω_{ϕ} is as follows. First construct an invariant state $\tilde{\omega}^{\epsilon}_{\phi}$ by the averaging procedure of 4.2.5. and then note that from 4.2.6. we have

$$S(\beta H^{\phi}\,;\,\tilde{\omega}^{\epsilon}_{\phi}) = S(\beta H^{\phi}\,;\,\omega^{\epsilon}_{\phi}) \geqslant C(\epsilon)\left[P(\beta,\phi) - \epsilon\right].$$

But now $\{\tilde{\omega}^{\epsilon}_{\phi}\,;\,\epsilon > 0\}$ is a directed set of invariant states and because the set $E_{\Omega}(R^{\nu})$ is compact in the weak* topology the state ω_{ϕ} defined by the weak* limit

$$\omega_{\phi} = \limsup_{\epsilon \to 0} \tilde{\omega}^{\epsilon}_{\phi}$$

must be an invariant state. It follows however from the upper semi-continuity of the conditional entropy and the criterion of 3.1.5. that

$$S(\beta H^{\phi}; \omega_{\phi}) \geq \limsup_{\epsilon \to 0} S(\beta H^{\phi}; \tilde{\omega}_{\phi}^{\epsilon})$$

$$\geq P(\beta, \phi) \; .$$

Hence from the first inequality of the proof we have

$$S(\beta H^{\phi}; \omega_{\phi}) = P(\beta, \phi)$$

and ω_{ϕ} attains the required supremum.

Secondly let us consider the solution of the above problem of finding periodic states for which the conditional entropy closely approximates the thermodynamic pressure.

As a consequence of the results of chapter 2 we can find a parallelepiped Λ_L centred at the origin with edges of length $L^{(1)}, L^{(2)}, \ldots, L^{(\nu)}$ parallel to the axes, such that

$$\frac{1}{V(\Lambda_L)} \log \mathrm{Tr}_{\mathcal{H}(\Lambda_L)} \left(e^{-\beta H^{\phi}_{\infty, \Lambda_L}}\right) \geq P(\beta, \phi) - \epsilon \; .$$

The suffix ∞ indicates that we take the Hamiltonian whose kinetic energy is defined with repulsive boundary conditions ; the interaction energy is of course defined with the aid of the potential ϕ. We denote by Λ_{L+r} the parallelepiped centred at the origin with the same orientation as Λ_L but with edges of length $L^{(1)}+r, \ldots, L^{(\nu)}+r$ where r is the range of ϕ and by Λ_s the complement $\Lambda_{L+r} \setminus \Lambda_L$ of Λ_L in Λ_{L+r} The Hilbert space $\mathcal{H}(\Lambda_{L+r})$ has then the tensor product decomposition

$$\mathcal{H}(\Lambda_{L+r}) = \mathcal{H}(\Lambda_L) \otimes \mathcal{H}(\Lambda_s) \, .$$

Now define a density matrix $\rho_{\Lambda_{L+r}}$ on this space by

$$\rho_{\Lambda_{L+r}} = \rho_{\Lambda_L} \otimes \rho_{\Lambda_s}$$

with the choices

$$\rho_{\Lambda_L} = e^{-\beta H^{\phi}_{\infty, \Lambda_L}} / \mathrm{Tr}_{\mathcal{H}(\Lambda_L)} \left(e^{-\beta H^{\phi}_{\infty, \Lambda_L}}\right)$$

$$\rho_{\Lambda_s} = E_0(\Lambda_s)$$

where $E_0(\Lambda_s)$ is the projector on the one-dimensional, zero-particle, subspace $\mathcal{H}^{(0)}(\Lambda_s)$ of $\mathcal{H}(\Lambda_s)$. Physically the density matrix $\rho_{\Lambda_{L+r}}$ represents a state of the finite system where particles are distributed with the Gibbsian equilibrium probabilities in Λ_L corresponding to the Hamiltonian $H^{\phi}_{\infty, \Lambda_L}$ and the probability of finding a particle in Λ_s is zero.

The density matrices $\rho_{\Lambda_{L+r}}$ will be used to construct a periodic state with period $L^{(1)}+r, \ldots, L^{(\nu)}+r$. Before we embark on this construction let us however consider the values of the local energy density in the state given by the above density matrix and in particular let us check that these values are finite.

Recall that theenergy densities $\hat{H}^{\phi}_{\Lambda}$ that we have to consider are constructed in the manner of chapter 3 but using a Hamiltonian $H^{\phi}_{0,\Lambda}$ which corresponds to $\sigma=0$ boundary conditions and potential ϕ. However this family of Hamiltonians has the property of 4.2.2. that for each state ω over $\mathfrak{A}_{\Lambda}$

$$\hat{H}^{\phi}_{\Lambda}(\omega) \geq \hat{H}^{\phi}_{\Lambda'}(\omega)$$

whenever $\Lambda' \subset \Lambda$. Thus to check that the energy densities of the normal state ω over $\mathfrak{A}_{\Lambda_{L+r}}$ given by the density matrix $\rho_{\Lambda_{L+r}}$ are finite we need only check that $\hat{H}^{\phi}_{\Lambda_{L+r}}$ is finite. But the eignevectors of $\rho_{\Lambda_{L+r}}$ are clearly of the form $\psi^{\rho}_{\Lambda_{L+r}} = \psi^{H}_{\Lambda_L} \otimes \psi^{0}_{\Lambda_S}$ where $\psi^{H}_{\Lambda_L}$ are the eigenvectors of $H^{\phi}_{\infty,\Lambda_L}$ on $\mathcal{H}(\Lambda_L)$ and $\psi^{0}_{\Lambda_S}$ is a fixed vector in the zero-particle space $\mathcal{H}^{(0)}(\Lambda_S)$. But each $\psi^{\rho}_{\Lambda_{L+r}}$ is then such that

$$\psi^{\rho}_{\Lambda_{L+r}}(X) = 0 \qquad X \cap \Lambda_S \neq \emptyset$$

and $\psi^{\rho}_{\Lambda_{L+r}}$ also vanishes on the boundary between Λ_L and Λ_S. Thus $\psi^{\rho}_{\Lambda_{L+r}}$ is continuous across this boundary. It is now straightforward to check that each of these eigenvectors is in the domain of the form associated with $H^{\phi}_{0,\Lambda_{L+r}}$ and is in fact in the domain of the form associated with $H^{\phi}_{\infty,\Lambda_{L+r}}$. However these forms coincide on their common domain and one consequently calculates that

$$\begin{aligned}\hat{H}^{\phi}_{\Lambda_{L+r}}(\omega) &= \mathrm{Tr}_{\mathcal{H}(\Lambda_L)}\left(e^{-\beta H^{\phi}_{\infty,\Lambda_L}} H^{\phi}_{\infty,\Lambda_L}\right) / \mathrm{Tr}_{\mathcal{H}(\Lambda_L)}\left(e^{-\beta H^{\phi}_{\infty,\Lambda_L}}\right) \\ &= -\frac{\partial}{\partial\beta} \log \mathrm{Tr}_{\mathcal{H}(\Lambda_L)}\left(e^{-\beta H^{\phi}_{\infty,\Lambda_L}}\right) \\ &< +\infty .\end{aligned}$$

(The finiteness is inferable from the convexity and boundedness conditions derived in chapter 2).

Note also that the entropy of the state ω is given by

$$S_{\Lambda_{L+r}}(\omega) = -\,\mathrm{Tr}_{\mathcal{H}(\Lambda_L)}\left(e^{-\beta H^{\phi}_{\infty,\Lambda_L}} H^{\phi}_{\infty,\Lambda_L}\right) \Big/ \mathrm{Tr}_{\mathcal{H}(\Lambda_L)}\left(e^{-\beta H^{\phi}_{\infty,\Lambda_L}}\right)$$
$$+ \log \mathrm{Tr}_{\mathcal{H}(\Lambda_L)}\left(e^{-\beta H^{\phi}_{\infty,\Lambda_L}}\right)$$

and hence

$$\frac{S_{\Lambda_{L+r}}(\beta H^{\phi};\omega)}{V(\Lambda_{L+r})} = \frac{V(\Lambda_L)}{V(\Lambda_{L+r})}\,\frac{1}{V(\Lambda_L)}\,\log \mathrm{Tr}_{\mathcal{H}(\Lambda_L)}\left(e^{-\beta H^{\phi}_{\infty,\Lambda_L}}\right)$$
$$\geq \frac{V(\Lambda_L)}{V(\Lambda_{L+r})}\left[\,P(\beta,\phi) - \epsilon\,\right] .$$

Let us now return to the construction of a periodic state over $\mathfrak{A}$. We define a locally normal state by the following prescription. If $a_n = (n^{(1)}(L^{(1)}+r), \ldots, n^{(\nu)}(L^{(\nu)}+r))$ with $n^{(1)},\ldots,n^{(\nu)}$ integers and $\Lambda_n = \Lambda_{L+r} + a_n$ we set

$$\rho_{\Lambda_n} = \tau_{a_n}\rho_{\Lambda_{L+r}} .$$

If

$$\Gamma_N = \bigcup_{n\in N} \Lambda_n$$

where N is a finite set of indices we set

$$\rho_{\Gamma_N} = \bigotimes_{n\in N} \rho_{\Lambda_n} .$$

Finally if $\Lambda\subset R^{\nu}$ we can always find a Γ_N such that $\Lambda\subset\Gamma_N$ and then we set

$$\rho_\Lambda = \mathrm{Tr}_{\mathcal{H}(\Gamma_N\setminus\Lambda)}(\rho_{\Gamma_N}) .$$

This construction gives us a family of compatible density matrices $\{\rho_\Lambda\}$ which define a state over $\mathfrak{A}$ with period $L^{(1)}+r, \ldots, L^{(\nu)}+r$. Let ω denote this state.

Now in the state ω there are particles distributed in cells Λ_L which are spaced a distance r apart. Thus the particles in different cells do not have any mutual interaction because the range of ϕ is exactly r. Hence it follows that the mean conditional entropy of the state ω is exactly the average conditional entropy for each cell of the lattice we have constructed. But then from the above calculation we have

$$S(\beta H^{\phi};\omega) \geqslant \frac{V(\Lambda_L)}{V(\Lambda_{L+r})}\left[P(\beta,\phi) - \epsilon \right]$$

and ω fulfils the condition of the required state ω^{ϵ}_{ϕ} because as $\epsilon \to 0$ we must have $L^{(\omega)} \to \infty$ and hence

$$V(\Lambda_L)/V(\Lambda_{L+r}) \longrightarrow 1 \text{ .}$$

Thus piecing together the above elements we have a complete proof of the proposition.

Note that in the above construction of the state it was absolutely essential that we used the Hamiltonian $H^{\phi}_{\infty,\Lambda}$ to construct the density matrices. Any other choice of boundary conditions in this construction would have given the value $+\infty$ to the mean energy. Using the notation of chapter 2 what we in fact proved was that

$$P^{\circ}(\beta,\phi) \geqslant \sup_{\omega\in E_{\mathfrak{a}}(R^{\nu})} S(\beta H^{\phi};\omega) \geqslant P^{\infty}(\beta,\phi)$$

and the desired result then follows only because we have already established the equality of the left and right hand members of this inequality.

§4 - EQUILIBRIUM STATES

4.4.1. We began our discussion of the variational principle in chapter 3 by considering a finite system and we pointed out that the (unique) density matrix for which the supremum is attained for the finite system

$$\rho^{*}_{\Lambda} = e^{-\beta H_{\Lambda}} / \mathrm{Tr}_{\mathcal{H}(\Lambda)}\left(e^{-\beta H_{\Lambda}}\right)$$

is interpreted as the equilibrium state of the system at inverse temperature β and with the (grand canonical) Hamiltonian H_{Λ}.

In the last section we showed that a variational principle can be established for the infinite, or thermodynamic, system and it is natural to define the states for which the supremum is attained as the translationally invariant thermodynamic equilibrium states. Of course this is not the only possible definition of equilibrium of all reasonable definitions. We will comment a little on this problem, but will principally point out properties of the states for which the supremum is attained which follow from the foregoing material and other well known results.

4.4.2. The set of translationally invariant thermodynamic equilibrium states at inverse temperature β and with interaction ϕ is defined by

$$\Delta(\beta,\phi) = \{ \omega \, ; \, \omega \in L(R^\nu) \, , P(\beta,\phi) = S(\omega) - \beta \hat{H}^\phi(\omega) \} .$$

First note that for $\beta > 0$ and ϕ positive, decreasing, and of finite range the result of 4.3.2. actually shows that $\Delta(\beta,\phi)$ is non-empty. Explicitly we can deduce from the weak* compactness of $E_{\mathfrak{A}}(R^\nu)$ and the affinity and upper semi-continuity of the conditional entropy that there are states $\omega \in E_{\mathfrak{A}}(R^\nu)$ which attain the maximum in the second variational principle of 4.3.2. ; this follows from 3.1.5. However we have then proved in 4.3.2. that these states are in $D(H^\phi) \subset L(R^\nu)$.

Now as $\Delta(\beta,\phi)$ is non-empty there are two cases which arise which we discuss separately.

4.4.3. For certain values of (β,ϕ) it can happen that $\Delta(\beta,\phi)$ consists of one point ω . If this is the case then it follows form 3.1.5. that ω is an element of the set of extremal points $\mathcal{E}(E_{\mathfrak{A}}(R^\nu))$ of the convex set of translationally invariant states over the C* algebra $\mathfrak{A}$. It however follows from general considerations, which are described in the book of Ruelle, that the extremal translationally invariant states are physically interpretable as states which describe a pure invariant thermodynamic phase. This interpretation is based upon an examination of the fluctuations of space-averaged physical observables, e.g. the density ; all such fluctuations are small if, and only if, ω is an extremal invariant state.

4.4.4. Let us now assume that (β,ϕ) are such that $\Delta(\beta,\phi)$ consists of more than one point. In this case it immediately follows from the affinity of the conditional entropy that $\Delta(\beta,\phi)$ is a convex subset of $D(H^\phi)$. But it also follows from the weak* upper semi-continuity of $S(\beta H^\phi;\cdot)$ that $\Delta(\beta,\phi)$ is compact in the weak* topology. To deduce this we have simply to show that $\Delta(\beta,\phi)$ is weak* closed because the compactness then follows because $\Delta(\beta,\phi)$ is a subset of the weak* compact subset $E_{\mathfrak{A}}(R^\nu)$ (cf. 3.1.4.). But if ω_α is a directed set of states in $\Delta(\beta,\phi)$ with limit point ω in $E_{\alpha}(R^\nu)$ we have from 4.2.3. and the upper semi-continuity that

$$P(\beta,\phi) \geqslant S(\beta H^\phi;\omega) \geqslant \lim_{\alpha} \sup S(\beta H^\phi;\omega_\alpha) = P(\beta,\phi)$$

i.e. $\omega \in \Delta(\beta,\phi)$.

Next we note that this compactness property can be restated in terms of the locally uniform topology introduced in 4.1.7. In fact on $\Delta(\beta,\phi)$ the weak* topology and the locally uniform topology coincide and $\Delta(\beta,\phi)$ is compact in the latter topology. Both these properties follow because the first estimate made in 4.3.2. shows that, for a suitable choice of n , the set $\Delta(\beta,\phi)$ is a subset of

$$D_n(H^\phi) = \{\omega\,;\ \omega \in E_{\alpha}(R^\nu)\ ,\ \hat{H}^\phi(\omega) \leq n\}\ .$$

However on this latter set the locally uniform topology and the weak* topology coincide and the set is closed in both topologies ; this follows from 3.2.3. and 4.2.3. Thus as $\Delta(\beta,\phi)$ is compact in the weak* topology it is compact in the locally uniform topology.

The final interesting property of the set $\Delta(\beta,\phi)$ is that it is a Choquet simplex whose extremal points are extremal points of $E_{\alpha}(R^\nu)$. This means that to each state $\omega \in \Delta(\beta,\phi)$ there corresponds a unique measure μ_ω concentrated on the extremal points of $\Delta(\beta,\phi)$ such that

$$\omega = \int d\mu_\omega(\omega')\,\omega'\ .$$

Further the ω' are states which correspond to pure thermodynamic phases. The details of the analysis which leads to these results can be found in Ruelle's book.

The physical picture of this property is that if $\Delta(\beta,\phi)$ contains many points there is a possibility of many invariant thermodynamic phases but each equilibrium state has a unique decomposition as a mixture of pure thermodynamic phases which are described by the extremal points of $\Delta(\beta,\phi)$.

4.4.5. There is one remaining property of the thermodynamic equilibrium states which can be deduced from the variational principle and this involves their connection with the finite volume equilibrium states. We illustrate this connection with a particular example. Take $\omega \in \Delta(\beta,\phi)$ then

$$P(\beta+\beta', \phi) \geq S(\omega) - (\beta+\beta')\hat{H}^{\phi}(\omega)$$
$$= P(\beta,\phi) - \beta'\hat{H}^{\phi}(\omega) .$$

Thus the function $\beta' \to \beta' \hat{H}^{\phi}(\omega)$ is a tangent function to the graph of $\beta \to P(\beta,\phi)$ at the point $\beta' = \beta$. Now assume that P is differentiable at the point β then we must have

$$\hat{H}(\omega) = -\frac{\partial}{\partial\beta} P(\beta,\phi) .$$

On the other hand for the finite system

$$-\frac{\partial}{\partial\beta} P_{\Lambda}(\beta,\phi) = -\frac{\partial}{\partial\beta} \frac{1}{V(\Lambda)} \log \mathrm{Tr}_{\mathcal{H}(\Lambda)} (e^{-\beta H^{\phi}_{\Lambda}})$$
$$= \mathrm{Tr}_{\mathcal{H}(\Lambda)} \left(e^{-\beta H^{\phi}_{\Lambda}} \frac{H^{\phi}_{\Lambda}}{V(\Lambda)}\right) / \mathrm{Tr}_{\mathcal{H}(\Lambda)} (e^{-\beta H^{\phi}_{\Lambda}})$$

and the right hand side of this equation represents the value of the energy per unit volume in the finite volume equilibrium state. But as $P_{\Lambda} \to P$ we can conclude by a straightforward geometrical argument that

$$\hat{H}^{\phi}(\omega) = \lim_{\Lambda\to\infty} \mathrm{Tr}_{\mathcal{H}(\Lambda)} \left(e^{-\beta H^{\phi}_{\Lambda}} \frac{H^{\phi}_{\Lambda}}{V(\Lambda)}\right) / \mathrm{Tr}_{\mathcal{H}(\Lambda)} (e^{-\beta H^{\phi}_{\Lambda}}) .$$

Thus the mean energy of the thermodynamic equilibrium state is the limit of the mean energies of the finite volume equilibrium states.

The form of argument used in this example has been elaborated for systems of particles on a lattice, or for particles with hard cores, to give a detailed connection between the finite volume and the thermodynamic equilibrium states. In these special cases, however, extra information is provided because the results aof these lectures, the existence of the thermodynamic pressure and the variational principle, can be derived for large classes of interactions.

It would be of interest to obtain such a generalization for the case of point particles.

4.4.6. We conclude by emphasizing that the topological and measure-theoretic properties of the equilibrium states which we have just briefly described are completely independent of the particular C* algebra $\mathfrak{A}$ used in their construction. All these properties are finally phraseable in the locally uniform topology and hence the arbitrariness of the choice of $\mathfrak{A}$, mentioned in 4.1.8. is irrelevant. In fact we could have considered the convex space L of families of compatible density matrices $\{\rho_\Lambda\}$ equipped with the locally uniform topology, in the manner of 4.1.9., and derived all results without mention of $\mathfrak{A}$. The introduction of $\mathfrak{A}$ and the set of all states $E_{\mathfrak{A}}$ is just a convenient method of compactifying L in a manner reminiscent of the Stone-Cech method.

BIBLIOGRAPHY

The variational principle discussed in this chapter was first introduced in the framework of classical lattice systems by

D. Ruelle

Commun. Math. Phys. 5 324 (1967)

Various generalizations to more complex systems have been given, some of which are described in :

D. Ruelle

Statistical Mechanics, Benjamin (New York) 1969

The properties of equilibrium states which we have mentioned are discussed in detail in this latter reference.

In the construction of the variational principle we have in part followed :

S. Miracle-Sole and D.W. Robinson

Commun. Math. Phys. (to be published)

EXERCISES

1. a) Let $\mathcal{H}$ be a Hilbert space and N the convex space of density matrices on $\mathcal{H}$ equipped with the trace-norm topology. Prove that N is not locally compact unless $\mathcal{H}$ is finite-dimensional.

b) Let $\mathfrak{A}$ be a C* algebra on $\mathcal{H}$ which contains the compact operators and identify N as a subset of $E_{\mathfrak{A}}$ (3.1.6.). Prove that the weak* and uniform (trace-norm) topologies of $\mathfrak{A}$ coincide on N, i.e. if $\mathfrak{A}$ contains an identity $E_{\mathfrak{A}}$ is a compactification of N.

[Hint : Consult D.W. Robinson - Commun. Math. Phys. 19, 219 (1970), theorem 1]

2. a) Let L be the space of states $\{\rho_\Lambda\}$ equipped with the locally uniform topology in the manner of 4.1.9. Prove that L is not locally compact.

b) Let $\mathfrak{A}$ be the quasi-local C* algebra defined by the generating family $\{\mathcal{L}(\mathcal{H}(\Lambda))\}$ and identify L as a subset of $E_{\mathfrak{A}}$. Prove that the weak* topology and the locally uniform topology coincide on L, i.e. $E_{\mathfrak{A}}$ is a compactification of L.

c) Prove that L possesses extremal points $\mathcal{E}(L)$, that $\mathcal{E}(L) \subset \mathcal{E}(E_{\mathfrak{A}})$, that L is the closed (in the locally uniform topology) convex envelope of the set of its extremal points $\mathcal{E}(L)$, and that $E_{\mathfrak{A}}$ is the closed (in the weak* topology) convex envelope of the set $\mathcal{E}(L)$.

[Hint : Use the result of exercise 1 and the known properties of $E_{\mathfrak{A}}$].

3. Let ω_ρ be a locally normal state. Prove that

$$x \in R^\nu \longrightarrow \omega_\rho(\tau_x A) \qquad , \qquad A \in \mathfrak{A}$$

is continuous.

[Hint : It suffices to consider the case $A \in \mathfrak{A}_\Lambda$ for some Λ. Let Λ' be such that $\Lambda \subset \Lambda'$ and $\Lambda + x \subset \Lambda'$ and take E_ρ to be a finite range projector on $\mathcal{H}(\Lambda')$ which commutes with the density matrix $\rho_{\Lambda'}$ associated with ω_ρ. One then has

$$
\begin{aligned}
|\omega_\rho(A) - \omega_\rho(\tau_x A)| &= |\mathrm{Tr}_{\mathcal{H}(\Lambda')}(\rho_{\Lambda'}(A - U_{\Lambda,x} A U_{\Lambda,x}^{-1}))| \\
&\leq |\mathrm{Tr}_{\mathcal{H}(\Lambda')}(\rho_{\Lambda'} E_\rho (A - U_{\Lambda,x} A U_{\Lambda,x}^{-1}) E_\rho)| \\
&\quad + |\mathrm{Tr}_{\mathcal{H}(\Lambda')}(\rho_{\Lambda'} (1 - E_\rho)(A - U_{\Lambda,x} A U_{\Lambda,x}^{-1})(1 - E_\rho))| \\
&\leq 2\|A\| \, \|(1 - U_{\Lambda,x}^{-1}) E_\rho\| + 2\|A\| \, \mathrm{Tr}_{\mathcal{H}(\Lambda')}(\rho_{\Lambda'}(1 - E_\rho)).
\end{aligned}
$$

Beschaffenheit der Manuskripte

Die Manuskripte werden photomechanisch vervielfältigt; sie müssen daher in sauberer Schreibmaschinenschrift mit ausreichend großer Type geschrieben sein. Handschriftliche Formeln bitte nur mit schwarzer Tusche eintragen. Notwendige Korrekturen sind bei dem bereits geschriebenen Text entweder durch Überkleben des alten Textes vorzunehmen oder aber müssen die zu korrigierenden Stellen mit weißem Korrekturlack abgedeckt werden. Die reproduktionsfähigen Abbildungen (in Originalgröße) sollen in den Text eingeklebt werden. Falls das Manuskript oder Teile desselben neu geschrieben werden müssen, ist der Verlag bereit, dem Autor bei Erscheinen seines Bandes einen angemessenen Betrag zu zahlen. Die Autoren erhalten 50 Freiexemplare.

Zur Erreichung eines möglichst optimalen Reproduktionsergebnisses ist es erwünscht, daß bei der vorgesehenen Verkleinerung der Manuskripte der Text auf einer Seite in der Breite möglichst 18 cm und in der Höhe 26,5 cm nicht überschreitet. **Entsprechende Satzspiegelvordrucke werden vom Verlag gern auf Anforderung zur Verfügung gestellt.**

Manuskripte, in englischer, deutscher oder französischer Sprache abgefaßt, sind einzureichen bei: Springer-Verlag, 6900 Heidelberg, Postfach 1780.

Cette série a pour but de donner des informations rapides, de niveau élevé, sur des développements récents en physique, aussi bien dans la recherche que dans l'enseignement supérieur. On prévoit de publier.

1. des versions préliminaires de travaux originaux et de monographies
2. des cours spéciaux portant sur un domaine nouveau ou sur des aspects nouveaux de domaines classiques
3. des rapports de séminaires
4. des conférences faites lors de congrès ou de colloques

En outre il est prévu de publier dans cette série, si la demande le justifie, des rapports de séminaires et des cours multicopiés ailleurs mais déjà épuisés.

Dans l'intérêt d'une diffusion rapide, les contributions auront souvent un caractère provisoire; le cas échéant, les démonstrations ne seront données que dans les grandes lignes. Les travaux présentés pourront également paraître ailleurs. Une réserve suffisante d'exemplaires sera toujours disponible. En permettant aux personnes intéressées d'être informées plus rapidement, les éditeurs Springer espèrent, par cette série de «prépublications», rendre d'appréciables services aux instituts de physique. Les annonces dans les revues spécialisées, les inscriptions aux catalogues et les copyrights rendront plus facile aux bibliothèques la tâche de réunir une documentation complète.

Présentation des manuscrits

Les manuscrits, étant reproduits par procédé photomécanique, doivent être soigneusement dactylographiés type assez grand. Il est recommandé d'écrire à l'encre de Chine noire les formules non dactylographiées. Les corrections nécessaires doivent être effectuées soit par collage du nouveau texte sur l'ancien soit en recouvrant les endroits à corriger par du vernis correcteur blanc. Les illustrations; en dimension originale, préparées pour reproduction sont à insérer dans le texte. S'il s'avère nécessaire d'écrire de nouveau le manuscrit, soit complètement, soit en partie, la maison d'édition se déclare prête à verser à l'auteur, lors de la parution du volume, le montant des frais correspondants. Les auteurs recoivent 50 exemplaires gratuits.

Pour obtenir une reproduction optimale il est désirable que le texte dactylographié sur une page ne dépasse pas 26,5 cm en hauteur et 18 cm en largeur. **Sur demande la maison d'edition met à la disposition des auteurs du papier spécialement preparé.**

Les manuscrits en anglais, allemand ou français peuvent être adressés à Springer-Verlag, 6900 Heidelberg, Postfach 1780.